TRANSPORTATION INFRASTRUCTURE -
ROADS, HIGHWAYS, BRIDGES, AIRPORTS AND MASS TRANSIT

ROADWAY WORK ZONE ANALYSIS

GUIDANCE FOR DECISION-MAKERS AND ANALYSTS

TRANSPORTATION INFRASTRUCTURE - ROADS, HIGHWAYS, BRIDGES, AIRPORTS AND MASS TRANSIT

Additional books in this series can be found on Nova's website under the Series tab.

Additional E-books in this series can be found on Nova's website under the E-book tab.

TRANSPORTATION INFRASTRUCTURE - ROADS, HIGHWAYS, BRIDGES, AIRPORTS AND MASS TRANSIT

ROADWAY WORK ZONE ANALYSIS

GUIDANCE FOR DECISION-MAKERS AND ANALYSTS

HEATHER P. WINFIELD
EDITOR

nova publishers
New York

For permission to use material from this book please contact us:
Telephone 631-231-7269; Fax 631-231-8175
Web Site: http://www.novapublishers.com

Library of Congress Cataloging-in-Publication Data

ISBN: 978-1-63117-155-0

Published by Nova Science Publishers, Inc. † New York

CONTENTS

PREFACE

Work zone planning and management has become increasingly challenging because of increasing travel demand and an aging roadway network infrastructure facing both more frequent maintenance and major rehabilitation projects. Road-operating organizations across the nation are increasingly motivated to reduce congestion impacts resulting from road work, particularly on roadways where congestion is already a recurrent feature even before road work begins. This book provides a work zone analysis and work zone modeling and simulation from traffic analysis tools volumes.

Chapter 1 – This document is intended to provide guidance to decision-makers at agencies and jurisdictions considering the role of analytical tools in work zone planning and management. It is often unclear what kind of analytical approach may be of most value, particularly in light of complex data requirements and staff training. The decision to create an analytical capability to support decision making can be a significant investment, and deserves careful consideration. In the end, work zone analysis should never be used to make key decisions but instead developed as a trusted resource for understanding the potential mobility impacts and using this information to inform key decisions.

Chapter 2 – This document is the second volume in the FHWA Traffic Analysis Toolbox: Work Zone Analysis series. Whereas the first volume provides guidance to decision-makers at agencies and jurisdictions considering the role of analytical tools in work zone planning and management, this volume provides specific guidance to the analyst, researcher, or manager in charge of conducting a specific work zone analysis project or who has been charged with developing an overall work zone modeling program or approach. This volume includes numerous case study examples, discussion and analysis designed to provide the prospective work zone analyst with information pertaining to the selection of a transportation modeling approach (including the identification of opportunities for use, managing technical risk and examples) as well as specific project applications (including constructability, scheduling and transportation management plan design and evaluation).

In: Roadway Work Zone Analysis
Editor: Heather P. Winfield

ISBN: 978-1-63117-155-0

Chapter 1

TRAFFIC ANALYSIS TOOLBOX VOLUME VIII: WORK ZONE MODELING AND SIMULATION—A GUIDE FOR DECISION-MAKERS*

Matthew Hardy and Karl Wunderlich

ABSTRACT

This document is intended to provide guidance to decision-makers at agencies and jurisdictions considering the role of analytical tools in work zone planning and management. It is often unclear what kind of analytical approach may be of most value, particularly in light of complex data requirements and staff training. The decision to create an analytical capability to support decision making can be a significant investment, and deserves careful consideration. In the end, work zone analysis should never be used to make key decisions but instead developed as a trusted resource for understanding the potential mobility impacts and using this information to inform key decisions.

* This is an edited, reformatted and augmented version of the Federal Highway Administration publication, FHWA—HOP-08-029, dated July 2008.

FOREWORD

Work zone planning and management has become increasingly challenging because of increasing travel demand and an aging roadway network infrastructure facing both more frequent maintenance and major rehabilitation projects. These two factors have sharpened interest in analytical tools to assist in better understanding projected work zone mobility impacts. An understanding of projected mobility impacts is critical for two reasons. First, the work zone planner/manager must be able to consider mobility impacts in a complex balance of life-cycle costs, safety, environmental, and other impacts. Second, mobility impact measures are used to support the analysis of other impacts (e.g., environmental impacts).

This document is intended to provide guidance to decision-makers at agencies and jurisdictions considering the role of analytical tools in work zone planning and management. It is often unclear what kind of analytical approach may be of most value, particularly in light of complex data requirements and staff training. The decision to create an analytical capability to support decision making can be a significant investment, and deserves careful consideration. In the end, work zone analysis should never be used to *make* key decisions but instead developed as a trusted resource for understanding the potential mobility impacts and using this information to *inform* key decisions.

1.0. INTRODUCTION

Road-operating organizations across the nation are increasingly motivated to reduce congestion impacts resulting from road work, particularly on roadways where congestion is already a recurrent feature even before road work begins. Work zone planning and management has become increasingly challenging because of increasing travel demand and an aging roadway network infrastructure facing both more frequent maintenance and major rehabilitation projects. Increased travel demand and increased need for road work has sharpened interest in the application of analytical tools to assist in better understanding, among many considerations related to work zones, projected *mobility impacts* resulting from work zones.

At the same time that more demands are being placed on efficient work zone planning, the range of analytical tools have become more complex; data sources more abundant; and computing technology more powerful; all of which can be used to assist in the decision-making process. It is often unclear what kind of analytical approach may be of most value, particularly in light of complex data requirements and staff training. Thus, the decision to use analytical tools to support decision making can be a significant financial and intellectual investment, and deserves careful consideration. The purpose of this document is to provide high-level guidance to decision- makers on the applicability and use of a broad range of analytical tools for work zone planning and management. Successfully deployed, an analytical capability to assess and characterize mobility impacts can be invaluable assets in minimizing road user delays, reducing overall costs of traffic management, and keeping key stakeholders informed (including the public).

This guidance document is intended to provide the local decision-maker with a broad, fundamental understanding of how analytical tools can be used to support work zone decision making throughout an entire project life cycle. The guidance is rooted in an overall philosophy that "*one size does NOT fit all*" with respect to the best analytical approach, that is, no single tool or analytical approach is the right answer for all work zone analyses. Work zones are associated with a vast range of road work activities, ranging from multi-billion dollar mega-projects to emergency pothole repair. Further, *mobility impacts* can be an issue for projects on almost any type of facility ranging from low-volume rural roadways to highly congested urban freeway interchanges.

The organization of the document reflects this broad analytical philosophy for work zones and is organized as follows:

- **Section 2.0 Work Zone Impacts**—Identifies a range of work zone impacts that should be considered when analyzing work zones. Options for an analyst considering these impacts are presented, and the role of mobility impacts assessment in addressing these considerations is discussed.
- **Section 3.0 Decision-Making in the Project Life Cycle**—Presents the context for decision making throughout a typical project life cycle. This includes explicit recognition that decisions made in earlier stages of the project will impact and/or constrain decisions in later stages.
- **Section 4.0 Transportation Analysis Approaches**—Describes the classes of analytical tools available to support work zone mobility analysis. While specific software titles are identified in this section, the discussion is focused on *classes of tools*, particularly in terms of

trade offs between the geographic/temporal scope of analysis with the desired level of detail for the work zone analysis.

- **Section 5.0 Synthesis and Implementation**—Provides a synthesis looking across all of the issues detailed in previous sections. This section includes an analysis process and checklist that may prove helpful in structuring an effective approach to developing an analytical process to support work zone planning.

2.0. Work Zone Impacts

There are many considerations that must be taken into account when planning, designing and implementing a transportation management plan for work zones which FHWA discusses in the publications *Developing and Implementing Transportation Management Plans for Work Zones* and *Work Zone Impact Assessment: An Approach to Assess and Manage Work Zone Safety and Mobility Impacts of Road Projects* [1, *2*]. The work zone impacts discussed in these two documents can be summarized as follows:

- Safety Impacts (motorists and workers)
- Mobility Impacts
- Economic Considerations
- Environmental Concerns
- User Cost

The analysis of work zone impacts often requires the comparison of options that differ in fundamental ways with potentially differing impact in all six areas identified above. The availability and maturity of tools designed to produce quantitative estimates vary significantly when considering the range of work zone impacts. The broadest range of tools and analytical techniques available to an analyst focuses on work zone mobility impacts. Further, when the full range of mobility impacts can be quantitatively assessed, these impacts can be used in other analyses of safety, economic, environmental, and user cost impacts.

As an example, consider motorist safety impacts (an inclusive analysis will also consider worker safety as well). Some factors an analyst might want to assess include:

- Potential for increased crash rates on the section of road affected by construction.
- Traffic diverting from the main route having to travel on a facility which is designed to support a lower volume of traffic, which may result in a higher crash rate.
- The diversion route may be longer than the main route and involve more intersections. Since more vehicle-miles are traveled, more crashes may occur.

An analysis of mobility impacts can be helpful in supporting a safety analysis of all three factors. The duration of congested conditions in the section of road affected by construction can be estimated when mobility impacts are projected. If the tool includes an estimate of diverted traffic volumes, these estimates can be key elements of an analysis of increased vehicle-miles traveled on alternative (possibly lower standard) roadways. The analysis of mobility impacts does not complete the safety analysis, but provides key estimates that support a robust safety analysis. In this case, crash rates on key facility types would have to be estimated and combined with the results of the mobility analysis.

A broader discussion of work zone safety considerations is available in the report *Work Zone Impacts Assessment: An Approach to Assess and Manage Work Zone Safety and Mobility Impacts of Road Projects* [2].

Other key impacts are user delay and economic considerations. These considerations can encompass aggregate estimates of user delay as well as impacts to local businesses and costs to the local agency. Many of these costs are based upon results of a mobility impacts assessment using transportation models. For example, road user delay can be estimated from speed calculations prepared using the Highway Capacity Manual (HCM) or taken directly from a mesoscopic simulation model. These data are converted to travel time, multiplied by the number of affected vehicles, and then multiplied by a factor representing the financial value of travel time. This monetary value can then be used as part of the decision-making process as to whether it is cost-effective to use a specific work zone temporary traffic control strategy such as constructing a temporary bridge to reduce road user delay. In some cases the diversion of traffic away from businesses located along a facility with significant work zone-related delay can be an important local concern. Here, an estimate of diverted traffic volume can be paired with projected revenue generated by passing motorists to provide an informed estimate of these economic impacts. This type of economic analysis was conducted for

Louis Lake Road in Fremont County, Wyoming where long-term road closures were considered [3].

Clearly mobility impacts assessment is only one component that a decision-maker must consider in the complex balance of safety, economic, environmental, user cost, and other impacts [2]. However, a robust mobility analysis is often instrumental in the assessment of these other factors. Broader issues in work zone planning are examined in a range of other FHWA documents and resources including the following:

- FHWA Office of Operations, Work Zone Safety and Mobility Program: http://www.fhwa.dot.gov/workzones
- Work Zone and Traffic Analysis: http://www.ops.fhwa.dot.gov/wz/traffic_analysis/wzta.htm
- Work Zone Impacts Assessment: An Approach to Assess and Manage Work Zone Safety and Mobility Impacts of Road Projects: http://www.ops.fhwa.dot.gov/wz/resources/final_rule/wzi_guide/index.htm
- Work Zone Traffic Management http://www.ops.fhwa.dot.gov/wz/traffic_analysis/wztm.htm
- FHWA Economic Analysis Primer http://www.fhwa.dot.gov/infrastructure/asstmgmt/primer.cfm
- Manual of User Benefit Analysis for Highways (commonly referred to as the *Red Book*) https://bookstore.transportation.org/item_details.aspx?ID=5

3.0. Decision-Making in the Project Life Cycle

Analyzing and mitigating the mobility impacts of work zones are closely linked with the life cycle of the road work project which encompasses a complex decision-making process detailed in a related FHWA publication, *Work Zone Impacts Assessment: An Approach to Assess and Manage Work Zone Safety and Mobility Impacts of Road Projects* [2]. This report provides a detailed assessment of the many factors present in planning and managing work zones and provides a framework for assessing and managing the multitude of work zone impacts. The report provides a context for understanding the range of issues and decisions faced by work zone planners and managers. Each of these issues and decisions represents a potential opportunity for models to inform and strengthen overall work zone planning.

While the report does mention the use of analytical tools and describes the types of tools available, it does not provide a detailed description of the various approaches to modeling mobility impacts of work zones nor address comprehensive analytical approaches supporting decision making throughout the project life-cycle. This volume of the Traffic Analysis Toolbox is a natural follow-on to provide a more detailed discussion of these tools and approaches for decision-makers.

The *Work Zone Impacts Assessment* report presents a typical program delivery process that includes four key components: System Planning, Preliminary Engineering, Design, and Construction [2]. For the purpose of providing guidance on the role of work zone analysis within this larger process, a simplified framework specific to analyze work zone mobility impacts was developed that condenses the project delivery process components of Preliminary Engineering and Design into a single category. The reason for collapsing these two categories into one is that from the perspective of work zone analysis, the functions conducted during both are similar with no distinct boundary between the two. In fact, these two activities are often conducted in tandem with each other and any distinctions are primarily programmatic in nature (30 percent, 60 percent or 90 percent complete) rather than technical (e.g., selecting a certain type of construction method) and do not have a strong bearing on the selection and use of a work zone analysis tool. Thus, the work zone analysis framework is defined as containing three primary components: **1) System Planning**, **2) PE/Design**, and **3) Construction**.

However, more important to the discussion at hand is not specifically where in the program delivery process one is (Planning, PE/Design, Construction) but rather the types of decisions that need to be made. These decisions are represented by three inter-related decision types that drive the overall work zone decision-making process as a three-part *decision-making engine*:

- **Scheduling Decisions**—Decisions impacting when work zone activity will occur, ranging from the selection of time of day (e.g., the decision to perform work at night), to days of week (e.g., the decision to restrict work zone activity to weekends only), to time of year (e.g., a decision to work in summer only). For longer projects this includes how the work will be phased and staged along the roadway network, potentially over several years.

- **Application Decisions**—Decisions pertaining to the construction technique to be used within the work zone (e.g., a decision to use cast-in-place techniques rather than a pre- cast approach). There are numerous construction techniques that may be considered (see Resources, FHWA Office of Infrastructure available at the end of this report for links to more information). The selection of any one of these techniques will have different implications for work zone planning.
- **Transportation Management Plan Decisions**—Decisions that determine how traffic will be managed while work zones are in place. This includes issues of temporary traffic control (TTC—control strategies, traffic control devices), public information (PI—public awareness strategies, motorist information), and transportation operations (TO—demand management, work zone safety management). For example, developing a transportation management plan for a two-lane rural roadway may involve selecting between TTCs such as temporary traffic signals or a flagging crew to support two-way, one lane operations; notifying the motoring public of possible delays through a PI campaign; and finally developing a TO enforcement plan to ensure worker safety. For large-scale projects, a critical component may be the identification and implementation of a detour route. A more detailed discussion of developing transportation management plans (TMPs) is available in the publication *Developing and Implementing Transportation Management Plans for Work Zones* [1].

The decision-making engine concept is visually represented in Figure 1, with each decision type represented by one of the three circles. **Scheduling** is the circle on top denoted with an "S"; **Application** is the circle on the bottom left with an "A"; and **Transportation Management Plan** is the circle on the right with "TMP". Adjacent to each circle is a **smaller circle used to indicate a relative level of finality regarding the decisions within each category.** For example, in Figure 1, all of the decisions regarding the Application have been made indicating there is little, if any, room to make adjustments or refine those decisions. In contrast, the decision regarding TMP is shown approximately 25 percent complete indicating that many of these decisions have not yet been finalized, implying considerable flexibility potentially remaining in this area.

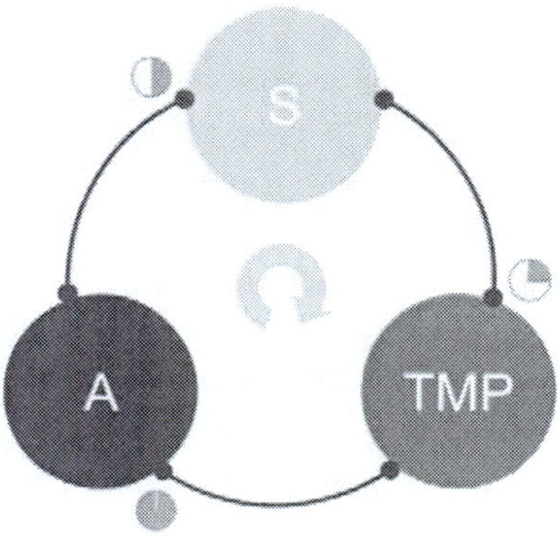

Figure 1. Work Zone Analysis Strategies Decision-Making Engine.

All of the three circles are connected indicating each decision type does not operate in isolation but is influenced by decisions made in other areas. Thus, a decision made about the application (e.g. cast-in-place concrete) may dictate the scheduling of the work (e.g. to work in warmer-weather months) which in turn impacts the transportation management plan that could be implemented. In the following sub-sections, the thickness and directionality of these connecting segments are used to graphically indicate the potential impact and direction of influence of decisions made in one area impacting or constraining decisions in other areas.

These decisions and their influences are highly dependent on when in the project life-cycle their relationships are examined. This natural dynamism in the process is represented by the central arrow in the center of the circle indicating rotational motion. As more decisions are made momentum is gained indicated by a darkening of the arrow. This arrow is a reminder that the decision-making engine is not static and is always moving forward and gaining momentum.

There are four key pieces of information the decision engine concept conveys:

1. The decision-making process is a dynamic one.
2. Decisions made in one category (scheduling, application or traffic management) affect decisions in other categories.
3. Decisions made in earlier stages of the project life-cycle will have an affect in later stages.

4. Once momentum is gained early in the planning process it becomes more difficult to deviate from that course of action later in the process.

These concepts are illustrated in Figure 2 where the size of the decision-making engine represents the momentum gained and as more and more decisions are made (represented by the two shaded arrows that grow larger as one moves from Planning to PE/Design to Construction), the amount of momentum grows larger (darkened central rotational arrow).

Figure 1 and Figure 2 present an overall conceptual framework which is used to better understand the concepts that are expanded upon in the following four subsections. The following subsections address in more detail the various decisions that have to be made during each stage (Planning, PE/Design and Construction), which components of the decision-making engine are more critical (scheduling, application and transportation management plan), and how decisions made during each stage affect available options later in the decision-making process.

3.1. Planning

The *Work Zone Impacts Assessment* guide defines systems planning as the stage of the larger program delivery process where planning for the future is carried out by identifying transportation system needs and deficiencies, developing and evaluating alternative improvement solutions, and compiling plans and programs for implementing the solutions [2]. While the *Work Zone Impacts Assessment* guide implies a much larger planning process examining an entire region, the planning process for implementing a work zone and accounting for its impacts has similar attributes, whether for a region or a specific project. During this process there are many questions about the impact of the work zone but the key decisions are often focused on the construction application to be used and scheduling of the construction. At this point in the process, the notion of a transportation management plan is somewhat limited since it will be heavily influenced by the scheduling and application. The Planning component of the decision-making process is shown in Figure 3.

As seen in Figure 3, each of the three decision categories is represented by a circle with an indistinct boundary indicating that although a general sense of feasible options may have been identified, all of the decisions to be made regarding each category have not been finalized. As the project moves from

Planning to PE/design many decisions regarding the application and scheduling will be made and the circles will become more distinct. The lines connecting the three categories indicate influences from the application technique impacting primarily the scheduling and the transportation management plan; and the dotted lines indicate that these interactions are considered at a high-level only. Similarly, the dotted line connecting scheduling and the transportation management plan indicate the two are accounted for but no formal decisions regarding each have been made. Finally, the central rotational arrow indicates that momentum is starting to build as decisions are made.

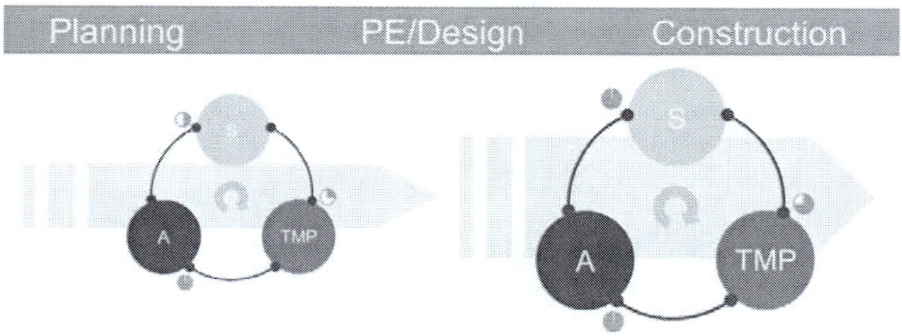

Figure 2. Decision-Making Engine within the Analysis Framework.

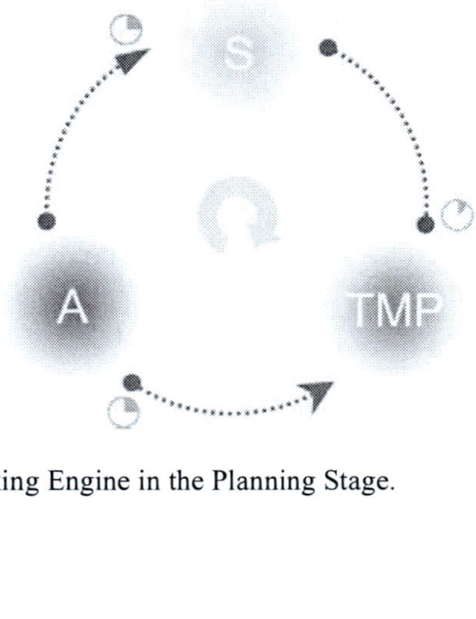

Figure 3. Decision-Making Engine in the Planning Stage.

Options in all three of the decision categories are influenced and constrained by other forces as well, e.g., funding availability, institutional policies or preferred contractual arrangements. For example, work zone scheduling may be staged to coincide with incremental funding available over multiple years. This may impact the application to be used (such as pre-cast concrete decking) which will in turn preclude some types of temporary traffic control options.

GOING TO THE SUN ROAD, GLACIER NATIONAL PARK

The application of work zone analysis in the planning stage is often used to assess the feasibility of broad project strategies relative to financial and institutional constraints. For example, when the Going to the Sun Road in Glacier National Park was in the planning stage for major reconstruction, an analytical approach was used to help determine the feasibility of conducting the work subject to a variety of constraints [3]:

- **Scheduling.** Road work limited to summer months in higher altitude sections, coinciding with the periods of highest travel demand. Overall staging also influenced by funding available in various years of the multi-year project.
- **Application**. Rugged terrain, as well as worker safety concerns, limited the use of certain construction applications.
- **Transportation Management Plan**. Given the critical nature of the road to the local economy, an agreement was made to maintain traffic flow at all times and to constrain delays from all work zones to 30 minutes or less.

The use of analytical tools early on in the process enabled decision-makers to explore the number and types of concurrent work zones along the scenic roadway that could be planned in each season over several years balancing funding constraints and mobility impacts. More information on the Going to the Sun Road can be found in the report *FLH-QuickZone Case Studies: The Application of FLH-QuickZone in Six Federal Lands Projects* [3].

3.2. PE/Design

The *Work Zone Impacts Assessment* guide separates PE/Design into two distinct categories within a larger construct of project design. The guide indicates that both stages of the project delivery process often occur in tandem without a clear-cut distinction between them [2]. During the PE/ Design stage some flexibility may remain in each category, however, key decisions influencing work zone activity are made in this stage—whether or not the potential mobility impacts have been actively considered or analyzed. For example, the primary application technique is often identified at this stage. As the project progresses in design, final drawings are being prepared influencing the feasible geometric plans for the work zone in general and transportation management plan in particular. Further, as the project moves forward in this stage, the overall project schedule takes on more tangible shape, typically identifying how long the project will take (duration) and when it will begin (start date). For example, as more decisions are made, prospective work zone analysts can better estimate how traffic will be impacted such as over the summer months or during the winter holiday season. As the decisions regarding applications and scheduling are firmed up, more emphasis is directed towards analyzing the impacts on traffic and developing the transportation management plan. The PE/Design component of the decision-making process is shown in Figure 4 below.

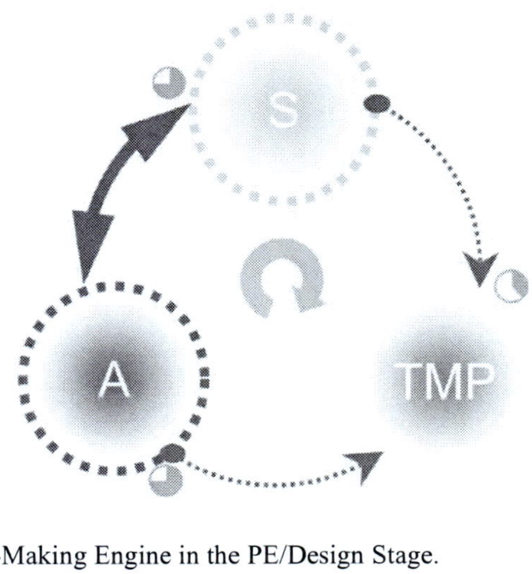

Figure 4. Decision-Making Engine in the PE/Design Stage.

As shown in Figure 4, each of the three decision areas is similar to those shown in Figure 3 with the difference being the dashed circle around the scheduling and application areas. Notice, too, that the central rotational arrow is darker in color. These circles indicate that decisions regarding these areas have been firmed up with more emphasis being transferred to the transportation management plans. Overall momentum is building, indicated by the central rotational arrow. The flexibility to go back and make significant changes to the overall approach declines as each decision impacting the work zone is made, not only because of the difficulty in altering a single aspect of the project, but because of the increasing inter-connection between decisions about the project.

The PE/Design stage represents a strong connection and dialog between the scheduling of the project and the application technique to be used. At this point, the decisions regarding the scheduling and application begin to constrain the types of transportation management plans available. By the end of this stage, the decisions regarding scheduling and application will have been made and the ability to change them difficult. Based upon the decisions being made about scheduling and application, a range of techniques for overall traffic control will begin to be explored. Decisions regarding the overall transportation management plan are still somewhat high-level at the beginning of the PE/Design stage but then firmed up as final designs are made. The decision to be made at this point includes placement of construction equipment and the use of temporary traffic control devices (flagger versus traffic signal). A more detailed discussion of developing a transportation management plan is available in the publication *Developing and Implementing Transportation Management Plans for Work Zones [1]*.

M-10 Lodge Freeway, Detroit, Michigan

A critical element that will impact overall work zone staging and operations during construction may be the application technique requiring a full closure or partial closure of the roadway. The reconstruction of portions of the M-10 Lodge Freeway in Detroit, Michigan was eventually conducted using a full closure. The decision to use this nontraditional technique was made during the PE/Design stage of the process, after consideration of four alternatives, and represented a constraint on the scheduling and transportation management plan.

This decision was supported, in part, by results from a simple analysis that indicated the abundance of alternate routes could handle the expected demand. The M-10 Lodge Freeway example demonstrates that the use of simple transportation analytical techniques during the PE/Design stage represents another opportunity where decisions can be further supported by a quantitative assessment. Further information about the M-10 Lodge Freeway project can be found in the case study *Full Road Closure for Work Zone Operations: A Case Study [4]*.

3.3. Construction

Construction is defined by the *Work Zone Impacts Assessment* report as the stage where the project is actually built. At this point, most decisions have been made regarding the application to be used though some modification may be made once construction begins [2]. Also, key decisions on scheduling have been firmed up except for some leeway in adjusting the schedule based upon flexible start dates built into a contract or the impact of unforeseen circumstances (construction equipment availability, weather, etc.). The transportation management plan has also been completed and will be implemented as part of the Construction stage. Though, the possibility of modifying the transportation management plan does exist based upon input from the contractor or other extenuating circumstances. In both instances (adjusting the schedule and modifying the transportation management plan) the need to reanalyze work zone mobility impacts may arise. The construction component of the decision-making process is shown in Figure 5 below.

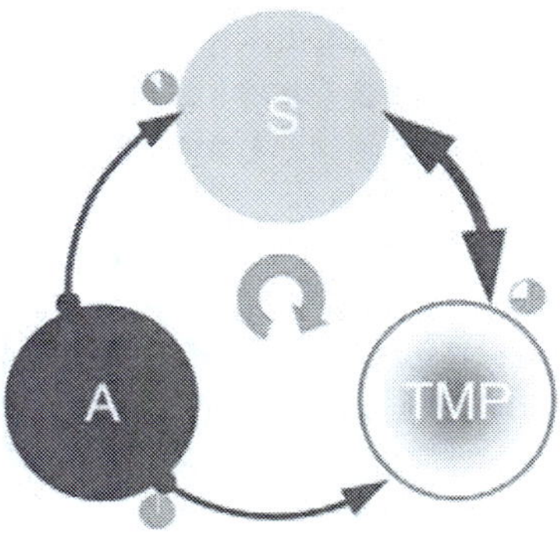

Figure 5. Decision-Making Engine in the Construction Stage.

As shown in Figure 5, most decisions regarding the application have been made along with decisions about the scheduling. The solid circle for the application component indicates that major changes to the construction technique are not likely to be made at this point. The shaded circle around the scheduling circle signifies there is still some flexibility available, but this is limited (for example, shifting the start of construction by a week rather than by a year). Where there is still room for change is with the transportation management plan. While the overall transportation management plan has been nearly complete by this point, unforeseen circumstances that were not accounted for in the Planning or PE/Design stage may reveal some needed modifications to the overall plan. At this point, overall momentum is at its greatest (indicated by the opaque central rotational arrow) and the ability for decision-makers to make changes is difficult (of course extreme decisions such as funding cuts may always be a possibility).

The Construction stage represents the strongest connections between scheduling and the transportation management plan. Planners, engineers and analysts in each area are working together to better ensure the impact to travelers is minimized. The decisions made here are being constrained by earlier decisions regarding application technique as well as outside policy forces. For example, the decision to use a certain type of high-strength concrete that must be placed during certain temperatures may force construction to begin earlier or later depending upon weather conditions. Also, policies limiting construction to non-holiday periods may impact the overall schedule start time.

WOODROW WILSON BRIDGE, NORTHERN VIRGINIA

In the case of I-495/Route 1 Construction Project in Northern Virginia, the construction contractor proposed the closure of a one block section of a local street to better facilitate the installation of various equipment that would have required intermittent lane closures over several months. Working with the Virginia DOT, the contractor's proposal was analyzed and the results indicated no significant mobility impacts at the local street level. In addition, traffic flow through the work zone was shown to improve. In the end, the recommendation was implemented. For more information see Appendix A in the *Work Zone Impacts Assessment report* [2].

Overall, during the Construction stage there is limited flexibility to make changes and what changes do have to be made will have to be supported by evidence and data so as not to impact the construction schedule or costs to the locality or the contractor. However, often the transportation management plan does need to be modified to accommodate unexpected conditions or unforeseen opportunities to improve work zone operations.

3.4. Decisions are Interdependent

The individual stages of the overall Work Zone Decision-Making Engine Process were discussed in the previous three sections. Figure 6 below combines all three individual stages into the overall work zone analysis framework. The three stages of the decision-making process are listed above the decision-making engine. Below it, represented by the two trapezoids, are the tradeoffs that are made between analysis opportunities and data requirements as the decision-making process evolves over time.

As seen in Figure 6, the evolution of the decision-making process occurs over time and begins with the Planning stage where various ideas, thoughts, and policies are discussed and analyzed at a somewhat high level. The momentum of the decision-making process is rather small and any changes made to either of the three components (application, staging, and transportation management plan) have limited effects on each other during this stage. However, decisions made in the Planning stage do impact later decisions in the PE/Design as well as Construction stage. These decisions represent constraints to the overall evolution of decision making and increase the overall momentum of the decision-making engine. During the Planning stage, the opportunity to use a variety of analysis tools (such as macro, meso, and micro which are discussed in the following section) is abundant and is primarily limited by agency resources (funding and institutional knowledge) or policies (agency requirements/guidelines to use a specific product). The opportunity to use an analytical tool to support decision making during the Planning stage is dependent on the availability of data. Often, detailed operational data is not available and can be a constraint on the ability to use certain types of models.

As the decision making process evolves over time from Planning to PE/Design to Construction, the analytical opportunities generally decline and data requirements for analysis increase as decisions become more detailed and exact. For example, the decisions that are made during the PE/Design stage regarding a construction technique (such as requiring the full closure of a heavily used bridge) have a direct impact on the ability of an analyst to assess the impact on the traveling public by requiring a specific type of tool (one that can account for detour routes). These decisions may also create future constraints in the Construction phase not only regarding the tool used, but the ability for the tool to yield results that are useful to the decision-makers themselves. In the end, the decisions that are made early on regarding the type of tool to use and the availability of data will limit the ability for analysts to provide results to decision-makers when the need arises. The Transportation Analysis Approaches section of this document discusses this aspect in more detail in relation to the analytical options available to practitioners.

4.0. TRANSPORTATION ANALYSIS APPROACHES

Transportation analysis involves the use of various methodologies and transportation software tools in order to better understand the effect a proposed alternative will have on the transportation network. FHWA maintains an extensive on-line resource devoted to the use of transportation modeling tools for analysis purposes called the Traffic Analysis Toolbox (TAT) (see the Resources section at the end of this document). The TAT organizes the available tools into seven categories: Sketch-Planning and Analytical/ Deterministic Tools (HCM-Based), Travel Demand Models, Traffic Signal Optimization, Macroscopic Simulation, Mesoscopic Simulation, and Microscopic Simulation [5].

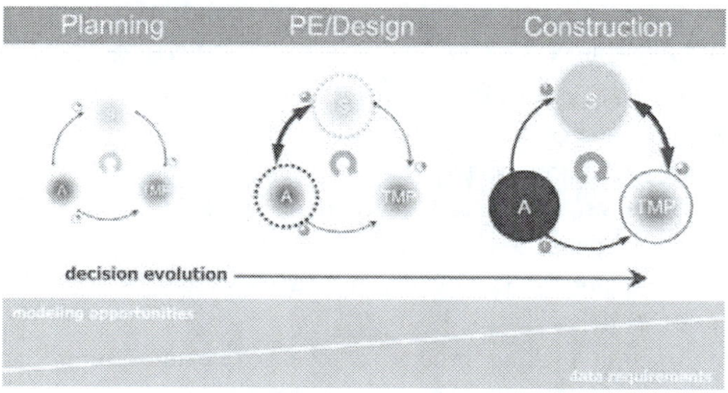

Figure 6. Work Zone Decision-Making Engine Process.

Each of the following sections on the analysis approaches includes a description of the basic concept that differentiates it from other approaches along with the strengths and weaknesses associated with the approach. This discussion takes place within the context of work zone analysis. A more detailed discussion of the application of various analytical approaches specifically for work zone analysis is provided in *Traffic Analysis Tools Volume IX: Work Zone Analysis: A Guide for Analysts*.

4.1. Sketch-Planning Tools and Analytical/Deterministic Tools (HCM Methodologies)

Sketch-Planning Tools and Analytical/Deterministic Tools (HCM Methodologies) (referred to here collectively as sketch-planning tools) are typically specialized models designed for a specific task or application such as work zone assessment or ITS analysis. These types of tools are in contrast to other general-purpose tools, such as travel demand models and microscopic simulation models which are targeted towards a much wider range of application, not just work zones analysis. Sketch- planning tools encompass a wide range of software including spreadsheet models developed by individuals and DOTs for specific projects or conditions, to more generalized delay estimation tools such as QUEWZ-98 and QuickZone. And, sketch-planning tools vary in complexity from analyzing only individual roadway segments to network-based analyses. Often, sketch-planning tools will be based upon simple queuing techniques or volume-to-capacity relationships from the Highway Capacity Manual. Although work zone specific sketch-planning tools have been available for many years, as a whole they continue to evolve and become more sophisticated in terms of features, use, and outputs.

The strength of sketch-planning tools rests upon their relative ease of use and ability to facilitate a rapid analysis. Typically, a sketch-planning tool will require fewer resources and less staff training to deploy than a mesoscopic or microscopic simulation model since they are simpler in terms of data requirements, calibration, and interpretation of the results. Regarding work zones, an analysis using a sketch-planning tool is normally quite rapid including both the input of the data and the model run time. This is important in cases where a decision needs to be made quickly or the agency desires a less resource-intensive analysis (e.g., for a project with a modest level of expected work zone impacts).

The strengths of these tools must be balanced with some of the weaknesses of them such as limited network complexity and a high-level analysis. For example, QUEWZ-98 only allows simple "pipeline" analysis without the ability to model cross-streets and detour routes. However, some of the more complex sketch-planning tools (e.g., QuickZone) do include the ability to model a detour route. The results of both of these models do not provide the detailed fidelity (level of accuracy and analysis) available in simulation models. In the end, the results from any sketch-planning tool will be relatively high-level (e.g., average or maximum queue) even though the input data is provided at a fairly detailed level (such as 15 minute traffic counts).

Lists of commonly used delay estimation tools are available at the following links:

- Sketch-Planning Tools: http://ops.fhwa.dot.gov/trafficanalysistools/tat_vol1/sectapp_a. htm#a1
- HCM Methodologies: http://ops.fhwa.dot.gov/trafficanalysistools/tat_vol1/sectapp_a. htm#a3
- Appendix B of *Work Zone Impacts Assessment*: http://www.ops.fhwa.dot.gov/wz/ resources/final_rule/wzi_guide/appb.htm

The use of delay estimation tools to analyze work zone impacts have been documented in a number of case study reports. These resources include the following links:

- FHWA *Work Zone Operations Best Practices Guidebook*: http://www.ops.fhwa.dot.gov/wz/practices/practices.htm
- *The Application of QuickZone in Eight Common Construction Projects*. Available from FHWA Office of Operations.
- *The Application of FLH-QuickZone in Six Federal Lands Projects*. Available from FHWA Federal Lands Highway Division

4.2. Travel Demand Models

Travel Demand Models are widely used today and were originally developed to model traffic in distinct transportation subnetworks, such as freeways, corridors (including freeways and parallel arterials), surface-street grid networks, and rural highways. The FHWA Traffic Analysis Toolbox describes travel demand models as being mathematical models that forecast future travel demand based on current conditions, and future projections of household and employment centers [5].

Travel demand models are traditionally thought of as the large regional planning models used by Regional Planning Commissions and Metropolitan Planning Organizations throughout the U.S. and were originally developed to determine the benefits and impacts of major highway improvements in metropolitan areas. To this end, a distinguishing feature of travel demand models is their geographic coverage. Generally speaking, travel demand models include an entire metropolitan area: a city, its suburbs, and the adjacent counties as required by law or need. The basic goal of these models is to forecast long-term future travel demand based on current conditions and projections of future household and employment characteristics.

For work zone analysis, one strength offered by travel demand models is their ability to predict area-wide traffic redistribution. For example, if an agency is considering closing an important urban freeway-to-freeway interchange for several months while it is rebuilt, a travel demand model could help evaluate the overall changes in total daily traffic volumes on various roadways throughout the region.

However, travel demand models have limited capabilities to accurately estimate changes in operational characteristics (such as speed, delay, and queuing) resulting from the implementation of operational strategies and changes (including the effects associated with roadwork construction). Because these models are prepared at a broad regional scale, they may lack accurate intersection turn volumes, so they may not be appropriate for detailed operational studies. They may model only one or two time periods (such as the AM peak hour or the daily average), which may not be sufficient for analyzing time-specific work zone traffic management strategies.

Some travel demand models contain only the total traffic volume on each link (combining the traffic volume from both travel directions). If this is the case, it may be difficult to adapt the model for analyzing routes or time periods that have unequal directional flows. For example, if the modeled volumes represent the Annual Average Daily Traffic, extensive re-working of the model may be required if the work zone analysis requires determining the directional peak hour flows into and out of a city's central business district.

Other limitations include their relative complexity and expense to build and maintain. However, these limitations must be balanced with the fact that these types of models can handle larger networks more efficiently than mesoscopic and microscopic simulation models. Another important consideration is that most major metropolitan areas already have a travel demand model constructed that could be used as a foundation for developing a transportation network for another model type.

A list of commonly used travel demand models is available at the FHWA Traffic Analysis Toolbox at http://ops.fhwa.dot.gov/trafficanalysistools/tat_vol1/sectapp_a.htm#a2.

4.3. Traffic Signal Optimization Tools

Traffic signal optimization tools are used to develop signal timing plans for isolated signal intersections, signalized arterial corridors, and signal networks. The FHWA Traffic Analysis Toolbox indicates that many of the available traffic signal optimization tools include the ability to conduct capacity calculations, cycle length determinations, splits optimizations, and coordination/ offset plans [5]. With respect to work zones, traffic signal optimization tools are useful when developing a signal plan for a temporary traffic signal or analyzing signal plans when a detour route directs traffic to an existing signalized arterial roadway. The primary limitation of these tools is their single focus. Traffic signal optimization tools are typically used to provide supplementary analysis when analyzing the overall mobility impacts of a work zone.

A list of commonly used traffic signal optimization tools is available at the FHWA Traffic Analysis Toolbox at http://ops.fhwa.dot.gov/trafficanalysistools/tat_vol1/sectapp_a.htm#a4.

4.4. Macroscopic Simulation Models

The FHWA Traffic Analysis Toolbox describes macroscopic simulation models as based upon the deterministic relationships of the flow, speed, and density of the traffic stream [5]. The simulation in a macroscopic model takes place on a section-by-section basis rather than by tracking individual vehicles. In other words, macroscopic tools treat traffic flows as an aggregated quantity; they do not *model the movement of individual vehicles on a network.*

In similar fashion to travel demand models, a characteristic feature of macroscopic models is their ability to cover large geographic areas. However, not all macroscopic simulation models include the ability to simulate a large network. The ability to model a large geographic area is useful when the work zone impacts may affect a larger corridor or region where large geographic impacts need to be better understood based upon a certain work zone design such as a full closure. Another characteristic of macroscopic simulation models pertaining to work zones is the run time. Because these models simulate aggregate flows, speeds, and density measures on each section of the network (rather than individual vehicles) they can be set up more quickly and run faster. This is useful where results need to be estimated quickly since these models generally require less time to run than travel demand models and other simulation models. The primary limitation of macroscopic models is their simple representation of traffic movement (e.g., no car following algorithms) which will limit the fidelity of the results.

A list of commonly used macroscopic simulation models is available at the FHWA Traffic Analysis Toolbox at http://ops.fhwa.dot.gov/trafficanalysistools/tat_vol1/sectapp_a.htm#a5.

4.5. Mesoscopic Simulation Models

Mesoscopic simulation models are the newest generation of traffic simulation modeling tools. These tools evolved from a need for an intermediate level of analysis. They provide more detail than both the travel demand models and macroscopic simulation models discussed previously, but not as much fidelity as the microscopic simulation tools discussed in the next section. Mesoscopic models tend to represent the relative flow of vehicles on a network link, but do not represent individual lanes on the link.

The strength of mesoscopic simulation models when analyzing work zones includes the ability to model both large geographic areas and corridors. In addition, diversion routes and signalized intersections can be modeled. In the case of a corridor with an Interstate and a signalized arterial road running through it, the diversion route onto the arterial could be more readily modeled using a mesoscopic simulation model. This type of analysis would not be easy to conduct using travel demand models or macroscopic simulation models, and can only be represented in a more limited fashion in some sketch-planning tools.

Mesoscopic simulation models do have a number of weaknesses. One is their limited ability to model detailed operational strategies such as complex signal control. Thus, if a work zone includes a number of signalized intersections, a different type of model (such as a microscopic simulation model or traffic signal optimization tool) may be a better choice. Another weakness of mesoscopic simulation models is the overall model complexity and data requirements necessary for accurate results. Mesoscopic models are an order of magnitude more complex than the most sophisticated sketch-planning tools, require similar amounts of data (albeit at a more granular level) as regional planning models, and similar resources (time, money, and knowledge) as travel demand models and microscopic simulation models.

Commonly used mesoscopic simulation tools for work zone analysis include the family of DYNASMART and DYNAMIT models as well as newly introduced hybrid/multi-scale models including Cube's Avenue and Caliper's TransModeler. Additional information on mesoscopic simulation models can be found at the following links:

- FHWA Traffic Analysis Tools—Mesoscopic Simulation Models: http://ops.fhwa.dot.gov/trafficanalysistools/tat_vol1/sectapp_a.htm#a6
- *Evacuation Management Operations Modeling Assessment: Transportation Modeling Inventory*. Available from the U.S. DOT RITA ITS Joint Program Office.

4.6. Microscopic Simulation Models

Microscopic simulation models were developed to accurately represent transportation systems at the individual vehicle level. They simulate the movement of individual vehicles based on car- following and lane-changing theories and other parameters. Microscopic simulation models update the positions and intentions of individual vehicles every second (or fraction of a second) as they move through a network. To account for the diversity of vehicles and driving styles that are encountered in real-world traffic, each vehicle is assigned a set of characteristics that influence the way it responds to the presence of other vehicles and to traffic control devices.

In the past, available computing power tended to limit the size of networks that could be modeled with microscopic simulation models. Recent advances in processor speed have reduced this problem but the desire to model larger areas and more complex travel behavior still exists. Today, the primary limitation of microscopic simulation models is the substantial amount of roadway geometry, traffic control, and traffic pattern data they require. Many transportation agencies currently use microscopic models in conjunction with travel demand models to better understand the impact of roadway geometry modifications on level of service and carrying capacity.

Today, microscopic simulation models are extensively used in a range of applications including evacuation planning and work zone impact analysis. These general-purpose models are effective in evaluating a wide range of scenarios including heavily congested conditions, complex geometric configurations, and system-level impacts of proposed transportation improvements that are beyond the limitations of other model types. While practical considerations may limit the geographic coverage of a microscopic simulation model, these models are useful in analyzing key bottlenecks on roadway segments and corridors where the movement of each individual vehicle needs to be represented to better understand the impact on roadway conditions.

A limitation of microscopic simulation models includes the inability to model large geographic areas. The reason is not a function of computing power, but the difficulty of calibrating a larger network that requires substantial data, resources, and technical expertise. The Michigan DOT has completed a microscopic simulation network of the entire Detroit Metropolitan Area freeway system which could be used for work zone analysis. More typical examples include smaller-scale analysis (in terms of geographic scale) for road projects such as the replacement of the Woodrow Wilson Bridge in Maryland and Virginia [2].

Commonly used microscopic simulation tools include CORSIM, VISSIM, and Paramics. Additional information on microscopic simulation tools can be found at the following links:

- FHWA Traffic Analysis Tools—Microscopic Simulation Models: http://ops.fhwa.dot.gov/trafficanalysistools/tat_vol1/sectapp_a.htm#a7
- *Traffic Analysis Toolbox Volume IV: Guidelines for Applying CORSIM Microsimulation Modeling Software:* http://ops.fhwa.dot.gov/trafficanalysistools/tat_vol4/vol4_guidelines. *pdf*

5.0. Synthesis and Implementation

This section pulls together the material discussed earlier in this document and addresses how to put these concepts into practice. First, some current tools from each of the seven categories are presented in a modeling spectrum. While it is useful to consider the categories of tools when developing an analytical approach for work zone planning and management, in the end specific models/tools are applied, not categories. Second, this section presents a checklist of conditions under which particular categories of tools are likely to be a valuable component of an analytical capability. The section ends with a summary of the motivation and benefits for developing and deploying an appropriate analytical capability in support of work zone decision making.

5.1. Work Zone Modeling Spectrum

As discussed in Section 4.0, there are many approaches available to analysts to address work zones, as illustrated in Figure 7 below. The spectrum described here includes seven of the many tools currently available that could be used to assess the impacts of roadway construction projects. Many other tools are available for use and are described in more detail at the Traffic Analysis Toolbox (TAT) website (see the Resources section for more detail). The tools shown here are used for comparison purposes and are placed on a continuum from simple to complex. Simpler tools include the categories of HCM and sketch-planning while the more complex tools include macro, meso, and microscopic simulation software.

There are many factors associated with selecting a specific methodology or tool which include functionality, level of detail (results and input data), time, training, and cost. Choosing a tool is generally a tradeoff among these five criteria. Functionality (the capability to represent specific work zone attributes) and level of detail (capability to quantitatively differentiate between alternatives) are two critical criteria because if the tool cannot analyze a specific situation or provide the necessary results to the precision or accuracy required, it would not be useful regardless of the cost, training, or development time applied. In some instances, such as a project currently in the construction phase, it may be critical that results are provided in a timely manner. The time required to assemble the required data and calibrate a simulation tool is generally much longer than the time and resources required to utilize a sketch-planning (delay estimation) tool. The need for timeliness must be balanced against the ability of the sketch-planning tool to provide a meaningful solution accurately reflecting the problem under study. Finally, the training and cost associated with a particular tool should be considered. Simulation tools often require a high level of expertise and training, skills which are possessed by relatively fewer individuals and can be costly to hire and retain. On the other hand, many of the sketch-planning and HCM tools are more accessible, can be mastered by a broad range of staff within a short time, and are inexpensive to purchase.

A critical element that will further impact these factors and tool selection is the availability of data. Regardless of the functionality desired or time available, the absence of data will influence both tool selection and the usefulness of analytical results.

5.2. Work Zone Analysis Process and Checklist

This subsection provides a high-level synthesis combining the concepts presented in Sections 2.0 through 4.0, and describes the various conditions under which the more common categories of tools is likely to be most helpful in supporting work zone planning and management: Sketch-Planning or Analytical/Deterministic Tools, Travel Demand/Macro Models, Mesoscopic Simulation Models, and Microscopic Simulation Models. A more detailed assessment is provided in the *report Traffic Analysis Tools IX: Work Zone Analysis: A Guide for Analysts* which includes concepts such as problem complexity, availability of resources and time, individual tool selection, fidelity of results, and multi-scale analysis approaches.

Not all conditions in an individual list need be present for the specific class of tools to be appropriate.

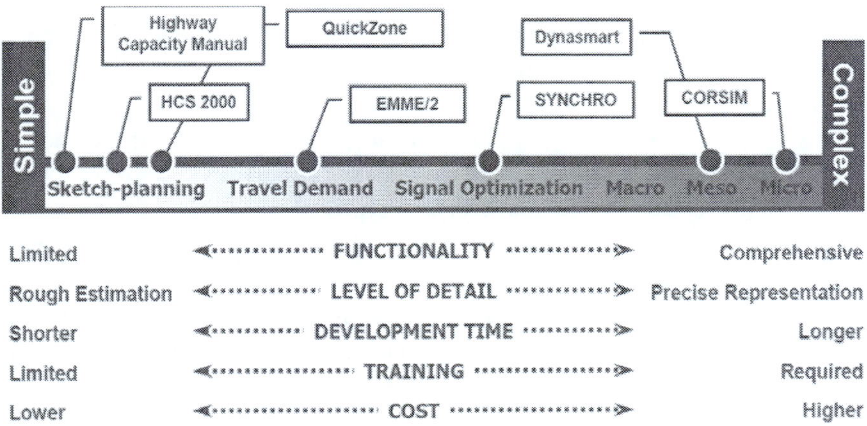

Figure 7. Work Zone Modeling Spectrum.

In what context is a *Sketch-Planning Tool or Analytical/Deterministic Tool (HCM Methodologies)* likely to be most useful?

- Work zone impacts are confined to a single facility; one or more work zones are located in a series along the facility (of any length).
- Resources for analysis are limited and rough estimation of congestion timing, extent, and intensity is sufficient to guide decision making.
- Mobility impacts must be rapidly assessed to support modifications to the transportation management plan, i.e., a quick response to a request for a change is required.
- Multi-season or multi-year projects wherein many (10+) potential phasing and staging sequences need to be assessed.
- A large number of smaller projects must be screened and ranked according to potential mobility impacts.
- No calibrated network analysis tool (meso or micro simulation) encompassing the area impacted by the planned work zone activity is available (or feasible to develop given resource and schedule constraints).

In what context is a *Travel Demand/Macro Model* likely to be most useful?

A calibrated, well-maintained travel demand or macro model is available encompassing the area impacted by the planned work zone activity.

- The time periods represented in the travel demand model coincide with those that are important for the work zone analysis.
- Work zone mobility impacts are expected over a large geographic area, particularly impacts on parallel and adjacent facilities, including transit networks.
- Accurate estimates of congestion timing, extent, and intensity are less critical than a need to identify the likelihood that congestion will form.
- Multiple concurrent and potentially interacting projects must be phased and staged across a broad geographic network with unknown mobility impacts.

In what context is a Mesoscopic Simulation Model likely to be most useful?

- A calibrated, well-maintained corridor model at the meso-scale is available encompassing the area impacted by the planned work zone activity.
- Work zone mobility impacts are expected on parallel facilities, including diversion impacts; key facilities generally feature uninterrupted flow (few signals or stop signs).
- Projects on parallel facilities must be conducted simultaneously with clearly interacting congestion and diversion impacts. Rates of diversion are expected to vary with changing congestion conditions in or near the work zone.
- Travel demand management and/or traveler information strategies are critical elements of the transportation management plan.
- Significant staff and data resources are available for model development and validation, including time-dependent vehicle counts at key bottlenecks in the network and dynamic end-to-end travel time estimates for the mainline and alternative routes at 15-minute intervals for recurring congestion.

In what context is a Microscopic Simulation Model likely to be most useful?

- A calibrated, well-maintained network for a traffic micro simulation is available encompassing the area impacted by the planned work zone activity.
- Critical performance measures are required at the lane level, or require the detailed assessment of alternative traffic control configurations (including signals) at interchanges, intersections, or lane shifts in the work zone.
- Diversion impacts are less critical than the detailed assessment of mainline traffic control.
- Significant staff and data resources are available for model development and validation, including network level descriptions of lane widths, turning movement bay lengths, traffic signal timing plans, and validation data including time-variant turning movement counts, vehicle counts by lane at critical bottlenecks, and end-to-end

travel time estimates for the mainline at 15-minute intervals for recurring congestion.

5.3. Summary

The results from analyzing transportation impacts can serve to improve decision making as well as improve overall understanding of the relationships between the many forces affecting work zone decision making: mobility, financial, environmental, safety, and user costs. Work zone analysis should never be used to *make* key decisions but instead developed as a trusted resource for understanding the potential mobility impacts and using this information to *inform* key decisions. The informative value of analysis used by decision-makes will directly relate to how well the analyst has considered both the *context for analysis* (decisions to be supported and relevant performance measures) and the *context for validation* (data and staff resources). The job of the work zone analyst extends beyond merely conducting an analysis and reporting results; but to provide decision-makers with a broader understanding that connects the findings of the analysis within the decision-making context. Placed in context, a well-summarized level of understanding can be provided to decision-makers and other staff working on the project, even if decision-makers do not have first-hand experience with the analytical approach.

RESOURCES

- **FHWA Office of Operations, Work Zone Safety and Mobility Program:** The FHWA Work Zone Mobility and Safety Program goal is to “make work zones work better” by providing transportation practitioners with high-quality products, tools, and information that can be used to help improve work zone management, and ultimately reduce congestion and crashes due to work zones. http://www.fhwa.dot.gov/workzones.
- **Work Zone and Traffic Analysis:** Understanding the anticipated type, severity, and extent of work zone impacts associated with various project alternatives facilitates the incorporation of appropriate mitigation measures and strategies in project programming, design, and in the development of effective transportation management plans (TMPs). This online resource provides a wealth of information related

to examples and use of work zone traffic analyses. http://www.ops.fhwa.dot.gov/wz/traffic_analysis/wzta.htm.

- **Work Zone Impacts Assessment: An Approach to Assess and Manage Work Zone Safety and Mobility Impacts of Road Projects:** This Guide is designed to help transportation agencies develop and/or update their own policies, processes, and procedures for assessing and managing the work zone impacts of their road projects throughout the different program delivery stages. http://www.ops.fhwa.dot.gov/wz/resources/final_rule/wzi_guide/index.htm.
- **Work Zone Traffic Management:** Online resource for managing traffic during construction to minimize traffic delays, maintain or improve motorist and worker safety, complete roadwork in a timely manner, and maintain access for businesses and residents. http://www.ops.fhwa.dot.gov/wz/traffic_analysis/wztm.htm.
- **FHWA Traffic Analysis Toolbox (TAT):** The Traffic Analysis Tools Program was formulated by FHWA in an attempt to strike a balance between efforts to develop new, improved tools in support of traffic operations analysis and efforts to facilitate the deployment and use of existing tools. The TAT provides a detailed description of applying traffic modeling tools http://ops.fhwa.dot.gov/trafficanalysistools/index.htm.
- ***FHWA Work Zone Operations Best Practices Guidebook:*** Keyword-searchable guide to nationwide best practices in areas such as modeling and impact assessment, planning and programming, contracting and bidding procedures, project design, public relations/outreach, and many others.
  http://www.ops.fhwa.dot.gov/wz/practices/practices.htm.
- ***The Application of QuickZone in Eight Common Construction Projects:*** This report summarizes the use of QuickZone for eight projects throughout North America. The analysis includes a listing of key observations that are important to how QuickZone was used for each project, a project overview, a discussion on network design for each case study, and an analysis of the results and how they were used Available from FHWA Office of Operations.
- ***The Application of FLH-QuickZone in Six Federal Lands Projects:*** This paper provides a summary of how FLH-QuickZone fits within a spectrum of other work zone delay estimation tools and how FLH-QuickZone was used for six FLH roadway construction projects. The paper concludes with observations on the challenge of addressing the

"soft cost" of roadway construction projects, through the use of FLH-QuickZone or with other analytical tools.
Available from FHWA Federal Lands Highway Division.

- ***Evacuation Management Operations Modeling Assessment: Transportation Modeling Inventory:*** This research documents more than 30 surface transportation modeling tools that have been applied or could be applied to evacuation modeling. Each tool represents a tradeoff between desired scope and analytical complexity, ranging from state-to-state coordination tools such as the Evacuation Traveler Information System (ETIS) to detailed traffic micro-simulation models such as the TSIS/CORSIM traffic simulation model. Available from the USDOT RITA ITS Joint Program Office.
- FHWA Office of Infrastructure: The Office of Infrastructure provides leadership, technical expertise, and program assistance in: Federal-Aid Highway Programs; Asset Management; Pavements; and Bridges to help sustain America's mobility.http://www.fhwa.dot.gov /infrastructure/.

REFERENCES

[1] Federal Highway Administration, *Developing and Implementing Transportation Management Plans for Work Zones*. 2005.

[2] Federal Highway Administration, *Work Zone Impacts Assessment: An Approach to Assess and Manage Work Zone Safety and Mobility Impacts of Road Projects*. 2006.

[3] Federal Lands Highway Division, *FLH-QuickZone Case Studies: The Application of FLH-QuickZone in Six Federal Lands Projects*. 2005.

[4] Federal Highway Administration, *Full Road Closure for Work Zone Operations: A Case Study*. 2004.

[5] Federal Highway Administration, *Traffic Analysis Toolbox, Volume 1: Traffic Analysis Tools Primer*. 2004.

In: Roadway Work Zone Analysis
Editor: Heather P. Winfield
ISBN: 978-1-63117-155-0

Chapter 2

TRAFFIC ANALYSIS TOOLS VOLUME IX: WORK ZONE MODELING AND SIMULATION—A GUIDE FOR ANALYSTS*

Matthew Hardy and Karl Wunderlich

ABSTRACT

This document is the second volume in the FHWA Traffic Analysis Toolbox: Work Zone Analysis series. Whereas the first volume provides guidance to decision-makers at agencies and jurisdictions considering the role of analytical tools in work zone planning and management, this volume provides specific guidance to the analyst, researcher, or manager in charge of conducting a specific work zone analysis project or who has been charged with developing an overall work zone modeling program or approach. This volume includes numerous case study examples, discussion and analysis designed to provide the prospective work zone analyst with information pertaining to the selection of a transportation modeling approach (including the identification of opportunities for use, managing technical risk and examples) as well as specific project applications (including constructability, scheduling and transportation management plan design and evaluation).

---

* This is an edited, reformatted and augmented version of the Federal Highway Administration, FHWA—HOP-09-001, dated March 2009.

FOREWORD

Work zone planning and management has become increasingly challenging because of increasing travel demand and an aging roadway network infrastructure facing both more frequent maintenance and major rehabilitation projects. These two factors have sharpened interest in analytical tools to assist in better understanding projected work zone mobility impacts. An understanding of projected mobility impacts is critical for two reasons. First, the work zone planner/manager must be able to consider mobility impacts in a complex balance of life-cycle costs, safety, environmental, and other impacts. Second, mobility impact measures are used to support the analysis of other impacts (e.g., environmental impacts).

This document is the second volume in the *FHWA Traffic Analysis Toolbox: Work Zone Analysis* series. Whereas the first volume provides guidance to decision-makers at agencies and jurisdictions considering the role of analytical tools in work zone planning and management, this volume provides specific guidance to the analyst, researcher, or manager in charge of conducting a specific work zone analysis project or who has been charged with developing an overall work zone modeling program or approach. This volume includes numerous case study examples, discussion and analysis designed to provide the prospective work zone analyst with information pertaining to the selection of a transportation modeling approach (including the identification of opportunities for use, managing technical risk and examples) as well as specific project applications (including constructability, scheduling and transportation management plan design and evaluation).

This document serves as Volume IX in the FHWA Traffic Analysis Toolbox. Preceding volumes in the toolbox include: Volume I: Traffic Analysis Tools Primer, Volume II: Decision Support Methodology for Selecting Traffic Analysis Tools, Volume III: Guidelines for Applying Traffic Microsimulation Modeling Software, Volume IV: Guidelines for Applying CORSIM Microsimulation Modeling Software, Volume V: Traffic Analysis Tools Case Studies - Benefits and Best Practices, Volume VI: Definition, Interpretation, and Calculation of Traffic Analysis Tools Measures of Effectiveness, Volume VII: Predicting Performance with Traffic Analysis Tools: Case Studies, and Volume VIII: Work Zone Modeling and Simulation—A Guide for Decision-Makers.

Paul Pisano,

Acting Director, Office of Transportation Operations

1.0. INTRODUCTION

This document is the second volume in the FHWA Traffic Analysis Toolbox: Work Zone Modeling and Simulation series. The intent of this volume *(TAT Volume IX)* is to provide guidance to the analyst, researcher, or manager in charge of conducting a specific work zone analysis project or who has been charged with developing an overall work zone modeling program or approach. This volume *is not* intended to be a specific how-to guide on using transportation analysis tools. This volume complements *Traffic Analysis Tools Volume VIII: Work Zone Modeling and Simulations—A Guide for Decision-Makers*, which was published with the intent to provide local decision-makers with a broad, fundamental understanding of how transportation analysis tools can be used to support work zone decision making throughout the complete project life cycle. These documents are rooted in an overall philosophy that "*one size does NOT fit all*" with respect to the best transportation modeling approach, that is, no single transportation analysis tool or strategic methodology is the right answer for all work zone analyses.

Similarly, it is important to recognize that using modeling and simulation tools for work zone impacts analysis may not be necessary for every project. The sophistication of the impacts analysis needs to be matched to the complexity and expected degree of impacts of the project. Work zone impacts analysis may involve a high-level, qualitative review for some projects, and a detailed, quantitative analysis using modeling and/or simulation tools for other projects. An agency should use modeling and simulation tools as helpful and appropriate for supporting its overall efforts to conduct work zone impacts analysis and making decisions to manage the impacts of projects.

The results from analyzing work zone impacts can serve to improve decision making as well as improve overall understanding of the relationships between the many forces affecting work zone decision making: mobility, financial, environmental, safety, and user costs. Critical to the work zone decision-making process is that a work zone analysis should never be used to *make* key decisions but instead developed as a trusted resource for understanding the potential mobility impacts and using this information to *inform* key decisions. The informative value of analysis is directly related to how well the analyst has considered both the *context for analysis* (decisions to be supported and relevant performance measures) and the *context for validation* (data and staff resources). The job of the work zone analyst extends beyond merely conducting an analysis and reporting results; instead the work zone analyst provides decision-makers with a broader understanding that connects the findings of the analysis within the decision-making context.

TAT Volume IX includes numerous case study examples, discussion and analysis designed to provide the prospective work zone analyst with information pertaining to:

1. Developing An Appropriate Analytical Approach
    a) Identifying opportunities for analysis throughout the project life-cycle
    b) Managing technical risk
    c) Incorporating lessons learned from case studies analyzing both simple and complex projects
2. Specific Project Applications
    b) Constructability—Constructability is defined as the optimum use of construction resources throughout the project life-cycle to achieve project objectives in a cost- effective manner. Opportunities to integrate mobility impacts assessment into constructability analyses, when identified and considered jointly, can often lead to innovative and high-payoff approaches minimizing mobility impacts and maximizing return on project investment.
    c) Scheduling—Transportation modeling approaches have the potential to provide analysts and decision makers with better information on how to time lane closures and other construction-related roadway capacity reducing measures to best reduce work zone mobility impacts. Scheduling activity can occur on a broad time scale, from assessing various alternative multi-year, multi-season approaches to scheduling hourly lane closures.
    d) Transportation Management Plan Design and Evaluation—Developing a Transportation Management Plan (TMP) is a critical component of many work zone projects. The use of transportation analysis tools can better evaluate the array of TMP options available to a transportation agency in order to identify the most effective options (or combination of options) with best potential to mitigate disruptions caused by roadway construction.

1.1. Transportation Modeling Approaches

The use of transportation analysis tools enables a decision-maker to better understand the impact a proposed alternative will have on the transportation network. The categorization of transportation modeling approaches found in *TAT Volume VIII and Volume IX* is based upon the FHWA Traffic Analysis Toolbox [2]. The FHWA TAT organizes the available tools into six categories: Sketch- Planning and Analytical/Deterministic Tools (HCM-Based), Travel Demand Models, Traffic Signal Optimization, Macroscopic Simulation, Mesoscopic Simulation, and Microscopic Simulation. A brief description of each category is provided here:

- **Sketch-Planning/HCM**—Specialized tools utilizing traffic count data and capacity analyses to predict transportation systems impact.
- **Travel Demand Models**—Mathematical models that forecast future regional travel demand based on current conditions, and future projections of household and employment centers.
- **Traffic Signal Optimization**—Optimization tools used to develop signal timing plans for isolated signal intersections, signalized arterial corridors, and signal networks.
- **Macroscopic Simulation**—Models based on deterministic relationships of the flow, speed, and density of the traffic stream. Macroscopic model simulations assess traffic conditions on a section-by-section basis rather than by tracking individual vehicles and treat traffic flows as an aggregated quantity; they do not model the movement of individual vehicles on a network.
- **Mesoscopic Simulation**—Intermediate approaches tailored for larger networks associated with Macroscopic and Travel Demand Models but preserving some aspects of the detail of Microscopic Simulation. Mesoscopic models estimate congestion effects based on the flow of vehicles across a link over time, but typically do not represent individual lanes on the link.
- **Microscopic Simulation**—These tools simulate the movement of individual vehicles based on car-following and lane-changing algorithms and other parameters associated with individual driver behavior. Microscopic Simulation models update the position of individual vehicles every second (or fraction of a second) as they move through a network by lane for every link the vehicle traverses.

TAT Volume VIII includes a detailed discussion of the various transportation modeling approaches available to analysts in order to analyze work zone impacts [1]. In Figure 1, these transportation modeling approaches are placed on a continuum from simple to complex. Above the continuum are specific examples of transportation analysis tools that have been used to conduct a work zone analysis (these examples do not represent all of the product choices). Simpler tools include the categories of HCM and sketch-planning while the more complex tools include macro, meso, and microscopic simulation software. The spectrum includes seven examples of commonly-used transportation analysis tools used to assess the impacts of roadway construction projects. A wider range of transportation analysis tools across the spectrum are described in more detail on the TAT website [2].

It is important to clarify some terminology used throughout the document. First, a transportation modeling approach is defined as one of the six model classifications identified in the work zone modeling spectrum (sketch planning/HCM through microscopic). Second, a transportation analysis tool is a specific computer program that is classified as one of the transportation modeling approaches. These tools are either freely available for download and use or through a commercial vendor. Finally, a work zone analysis is the process of an analyst using a transportation analysis tool to generate results for use by decision makers.

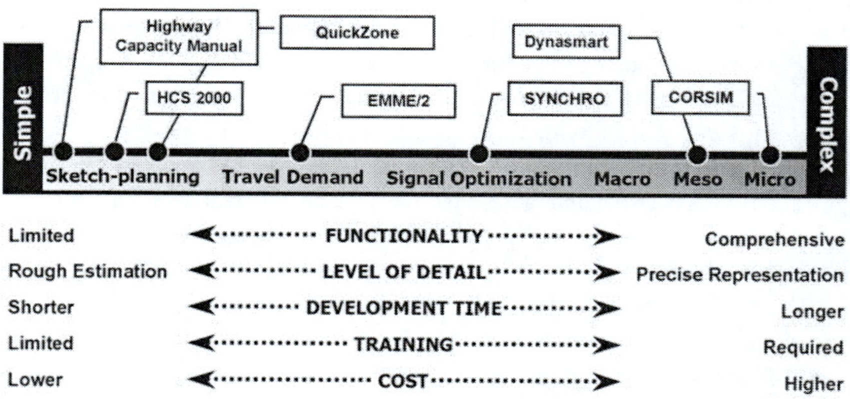

Figure 1. Work Zone Modeling.

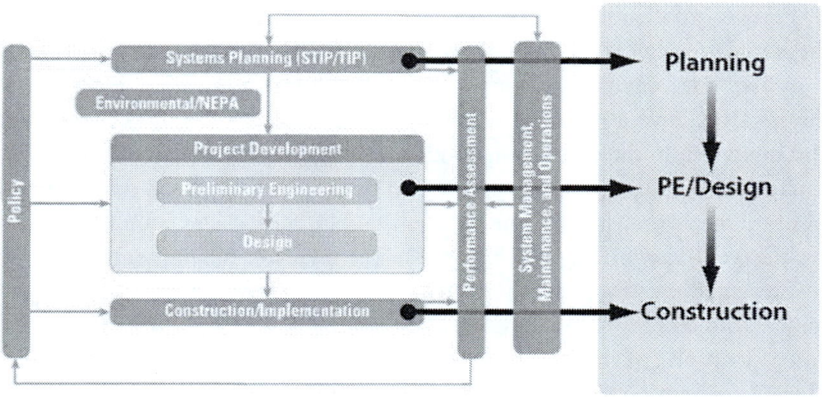

Figure 2. Work Zone Decision-Making Process.

Which transportation modeling approach is best suited to a particular work zone analysis? This document is intended to be a guide to assist analysts with answering this question, carefully considering the many factors associated with a specific transportation modeling approach or analysis tool. These factors include work zone characteristics, transportation management plan strategies, available data archives, agency resources and modeling capabilities, and work zone performance measures. Choosing a transportation modeling approach is generally a tradeoff among these five categories of factors. However, the key to successful work zone analysis does not begin or end with just transportation modeling approach selection, but rather in the successful integration of data and tools to provide a meaningful assessment of work zone impacts relevant to one or more key project decisions.

In essence, the sophistication of the analysis needs to be matched to the complexity of the project. For example, developing a microscopic simulation model would probably be "overkill" for a simple bridge replacement project on a minor rural highway. Similarly, a sketch planning tool alone would probably be insufficient for a comprehensive analysis of the impacts of a high-volume urban freeway-to-freeway interchange replacement project.

1.2. Work Zone Analysis Decision-Making

Figure 2 shows how the work zone decision making process, which consists of Planning, PE/ Design and Construction, was developed based upon the FHWA publication *Work Zone Impacts Assessment: An Approach to Assess and Manage Work Zone Safety and Mobility Impacts of Road Projects.*

While the decision-making process is important and transportation analysis tools have a potential role in all three stages of the project life-cycle, more important to the discussion at hand are the types of decisions that need to be made. These decisions are represented by three inter-related decision types that drive the overall work zone decision-making process as a three-part work zone *decision-making engine* (Figure 3). The decision-making engine concept is represented by three decision areas including **Scheduling** (circle on top denoted with an "S"); **Application** (circle on the bottom left with an "A"); and **Transportation Management Plan** (circle on the right with "TMP"). Adjacent to each circle is a smaller circle used to indicate a relative level of finality regarding the decisions within each category. For example, in Figure 3, all of the decisions regarding the Application have been made indicating there is little, if any, room to make adjustments or refine those decisions. In contrast, the decision regarding TMP is shown approximately 25 percent complete indicating that many of these decisions have not yet been finalized, implying considerable flexibility potentially remaining in this area.

All of the three circles are connected indicating each decision type does not operate in isolation but is influenced by decisions made in other areas. Thus, a decision made about the application (e.g., cast-in-place concrete) may dictate the scheduling of the work (e.g., to work in warmer-weather months) which in turn impacts the transportation management plan that could be implemented. In summary, the decision-making engine shown in Figure 3 conveys four key pieces of information:

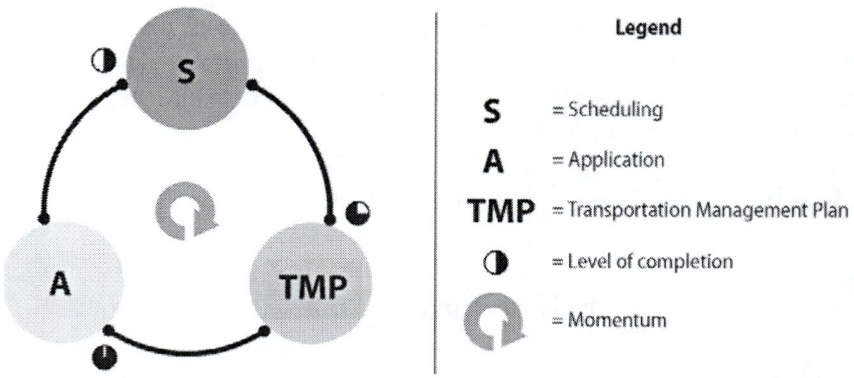

Figure 3. Work Zone Decision-Making Engine.

1. The decision making process is dynamic.
2. Decisions made in one category (scheduling, application, or traffic management) impacts decisions in other categories.
3. Decisions made in earlier stages of the project life-cycle will have an impact in later stages.
4. Once momentum is gained early in the planning process it becomes more difficult to deviate from that course of action later in the process.

1.3. Case Studies

A critical component for *TAT Volume IX* is the inclusion of case studies to demonstrate how different transportation agencies have applied transportation analysis tools for the purpose of work zone impacts assessment. The case studies appear in various elements of the document with varying levels of detail in their description. First, a brief summary of the case studies is provided in this section. Second, the case studies are used as specific examples throughout the document to highlight various aspects of the factors that go into developing a modeling approach (Section 2.0) or the process that a transportation agency has established to guide analysts in selecting a modeling approach (Section 3.0). Third, a summary of each case study is provided in the Case Studies section with references to additional reports and contacts for each case study.

A total of 17 case studies have been included in this volume (13 project applications and 4 strategic modeling approaches) and are shown in Figure 4 below. The 17 case studies have been categorized as either Project Applications or Strategic Modeling Approaches and are indicated by the diamond and fill, respectively, in the figure. The case studies that are included in this document were selected based upon access to the analysts and/or decision makers, availability of reports and data regarding the work zone application, geographic diversity, and project application (uniqueness). A summary of the categories and brief description are provided following the figure.

Project Applications—Project applications are specific examples of work zones associated with roadway construction projects that have used a transportation analysis tool to conduct a work zone analysis as part of the overall decision-making process.

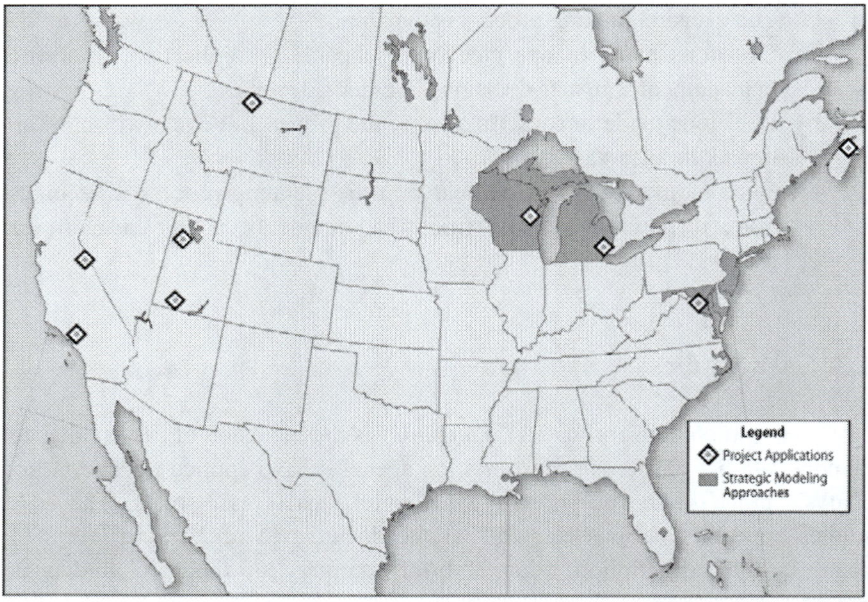

Figure 4. Case Study Locations.

1. *Caltrans I-15 Pavement Reconstruction Project (CA I-15)*—Application of an HCM-based constructability tool and mesoscopic simulation model during the design stage to address construction staging and mobility impacts.
2. *Glacier National Park: Going to the Sun Road Rehabilitation Project (GNP)*—Application of a sketch-planning tool in the planning stage to help assess scheduling and constructability issues surrounding a high-profile, multi-year construction project.
3. *Michigan DOT: Ambassador Gateway Bridge MOTSIM (MI AMB)*—Application of a microscopic simulation model during the final stages of design and beginning of construction in order to address maintenance of traffic issues.
4. *Michigan DOT: I-94 Rehab MOTSIM (MI I-94)*—Leveraged the Ambassador Gateway Bridge microscopic simulation model (see above) as part of the initial design stage in order to address constructability, staging, and maintenance of traffic.
5. *Michigan DOT: I-75 Trade Corridor MOTSIM (MI I-75)*—Leveraged and built upon the overall process developed as part of the Ambassador Gateway Bridge microscopic simulation model (see

above) to address mobility, constructability, and staging from the planning stage forward.

6. *Nova Scotia, Canada: Reeves Street (NS-Reeves)*—Application of a sketch planning tool for a simple work zone which includes a detour route.
7. *Utah DOT I-15 Reconstruction Design-Build Evaluation (UT I-15)*—Use of a travel demand model to determine the impact of selecting an innovative contracting technique (design-build) in terms of mobility, cost effectiveness, and safety.
8. *Wisconsin DOT Work Zone Signal Optimization*—Two examples of work zone signal optimization applications for work zones including a hypothetical case study used for training and a real-world application involving the Daniel Webster Hoan Memorial Bridge.
9. *Woodrow Wilson Bridge Reconstruction: Lane Closure Analysis (WWB-LCA)*—Application of an HCM-based tool to address contractor requests for additional lane closures during the construction stage.
10. *Woodrow Wilson Bridge Reconstruction: Roadway Operations Analysis (WWB-ROA)*— Application of a microscopic simulation tool during the construction stage to determine the optimal design of a traffic switch connecting old roadways to new alignments.
11. *Woodrow Wilson Bridge Reconstruction: Roadway Closure Analysis (WWB-RCA)*— Application of a sketch-planning tool during the construction stage to determine impact of significant lane closures and full closures for extended weekend work.
12. *Yosemite National Park: Yosemite Village Roadway Reconstruction (YOS)*—Application of a sketch-planning tool during the design stage to determine work zone staging and constructability.
13. *Zion National Park: Entrance Booth Reconstruction (ZION)*—Application of a sketch-planning tool during the design stage to determine construction scheduling.

Strategic Modeling Approaches—Strategic modeling approaches are examples of agencies setting up a systematic process to facilitate rapid assessment of work zone impacts and to ensure best practices when more complex modeling approaches are warranted.

1. *Maryland SHA Lane Closure Analysis Program (MD-LCAP)*—The Maryland State Highway Administration developed the Lane Closure

Analysis Program (LCAP) to support state traffic engineers with a structured method to analyze work zone impacts.

2. *Michigan DOT: Southeastern Michigan Simulation Network (MDOT-SEMSIN)*—Michigan DOT has developed a relatively complex microscopic simulation network and analysis process whereby decision-makers can leverage previous analyses for current work.
3. *New Jersey Turnpike Authority: Lane Closure Application (NJTA-LCA)*—The New Jersey Turnpike Authority established an approach whereby a simple GIS-based tool was developed to assist personnel in determining optimal timing of routine roadway maintenance work for the simplest of work zones.
4. *Wisconsin DOT: Transportation Management Plan Development Process (WisDOT)*—The Wisconsin DOT has established a formal Transportation Management Plan Development Process which includes a decision-tree on recommended tools to use based upon certain work zone characteristics. In addition, for the most high-profile work zones requiring detailed microscopic simulation analysis, they have also established a review process to ensure the proper use of these complex tools.

1.4. Document Organization

TAT Volume IX is structured around three broad categories of decisions associated with selecting and using a transportation modeling approach for better understanding work zone impacts on roadway construction projects: factors associated with conducting a work zone analysis; various strategic methodologies in addressing work zone analyses; and how to ultimately identify and develop a comprehensive transportation modeling approach for work zone analysis. Each of these three categories builds upon each other taking the reader from very detailed and finite elements associated with work zone analysis in general (do I have enough data or financial resources?) through the general concerns and questions that need to be addressed when ultimately deploying the transportation modeling approach.

The organization of the document reflects these three broad categories of considerations and is organized as follows:

- **Section 2.0 Work Zone Analysis Factors**—Provides a detailed discussion regarding five categories of work zone analysis factors that should be considered when deciding whether or not to conduct a work zone analysis and those data elements necessary to successfully conduct a work zone analysis. Section 2.0 includes a number of tables that provide a summary of the transportation modeling approach best suited based upon the factor being discussed. The summary tables are based upon the suitability classification system developed in *TAT Volume I* and are slightly modified as needed for each of the work zone analysis factors being discussed [2].

- **Section 3.0 Establishing a Strategic Methodology for Work Zone Analysis**—Presents three strategic methodologies to systematically incorporate transportation modeling into the overall work zone decision-making process.
- **Section 4.0 Identifying a Transportation Modeling Approach for Work Zone Analysis**— Provides a framework and process that can be used to select an overall transportation modeling approach appropriate for the given circumstances.
- **Section 5.0 Summary and Synthesis**—Provides a synthesis of the document.

2.0. WORK ZONE ANALYSIS FACTORS

The ultimate reason for developing and deploying a transportation modeling approach is to help make an informed decision regarding the implementation of one or more work zones. These key decisions can occur at any stage during the project life cycle including Planning, PE/Design, and Construction. For example, during the Construction stage, a contractor may propose a change to an approved lane closure plan in order to better utilize available time in the roadway by extending the overnight work hours into the morning commute. The analyst employed by the transportation agency will need to assess the merits of the proposed change and consider how to develop a feasible analytical assessment strategy by answering a number of questions related to the work zone[1]. First, how complex is the work zone area to be modeled? Is it an isolated element or part of a larger network? Second, what type of data is available to conduct an assessment such as traffic volumes? How will the proposed changes affect the transportation management plan? What resources are available to the analyst including staff resources, model and data availability, as well as the time available to actually conduct the analysis? Finally, what is the critical measure of system performance that will characterize mobility impacts of the proposed change: vehicle queue extent, vehicle delay, cumulative travel time? All of these considerations create constraints or opportunities for developing an effective analytical approach.

This example highlights an important issue for many transportation agencies. If the agency has already developed a robust analytical capability regarding a particular work zone, analyzing the contractor's proposal is relatively straight forward (plug the new alternative into an existing model). Conversely, an agency that needs to respond to the contractor's proposal quickly may not be readily able to conduct an analysis given the analytical capability does not exist and a work zone model has to be built from scratch. This could result in a rushed analysis that lacks accuracy or uses a tool that is too simple to analyze the proposal correctly. Therefore, many transportation agencies are beginning to think of work zone analyses in terms of an ongoing process that begins during project planning, is refined during preliminary engineering and final design, and is available for quick response during construction.

Regardless of project stage, a number of factors will ultimately influence the selection and use of a specific transportation analysis tool by an analyst. Figure 5 provides a graphical representation of the categories of factors influencing the selection of a work zone analysis approach.

- **Work Zone Characteristics**—Facility type, network form, and geographical scale of the potential work zone impact area.
- **Transportation Management Plan Strategies**—Temporary traffic control strategies, public information campaigns, and transportation operations.
- **Data**—Availability of data types (roadway characteristics, travel demand) as well as sources and quality of the data.
- **Agency Resources**—Institutional arrangements as well as the technical capability of the agency including modeling experience, model availability, and funding.
- **Work Zone Performance Measures**—The critical metrics used to differentiate alternatives and characterize impacts (e.g., maximum queue or additional travel time) as well as the precision of the results.

All of these factors have to be evaluated in order to select an appropriate work zone analysis approach. As represented by the black dotted lines, one category of factors may influence other categories. Thus, agency resources (in terms of budget constraints) may limit the quality and amount of data available which may affect the selection of a transportation modeling approach. What is important to note is that the connection between the work zone analysis approach and TMP is bidirectional indicating a work zone analysis approach is *influenced by* the TMP as well as *influencing development of* the TMP. The TMP cannot be treated as simply another factor in deciding which work zone analysis approach to select, but also to help guide the analyst in the selection of a work zone analysis approach as well.

The following five subsections will discuss each of these categories in more detail by providing a definition of the various terms and concepts, what is meant by each, and providing case study examples of how analysts have successfully addressed these factors in a work zone assessment.

2.1. Work Zone Characteristics

These sub-sections discuss how the physical attributes of the work zone itself, including the type of work zone, network form, and the geographic scale of the work zone impact area, shape and influence the modeling approach to be used.

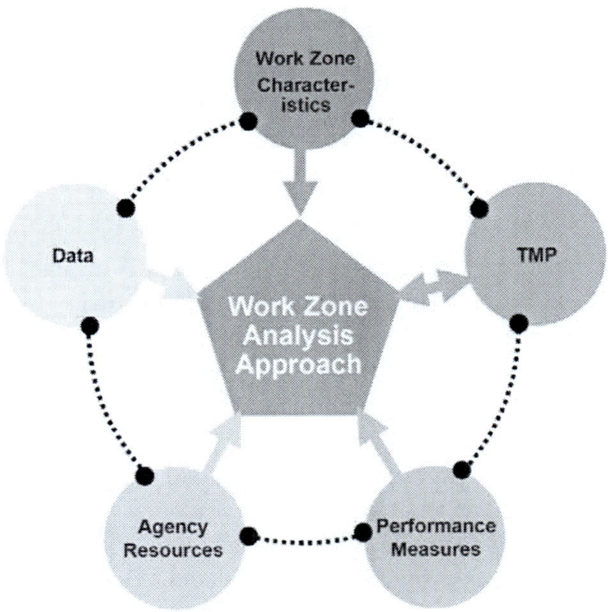

Figure 5. Work Zone Analysis Factors.

2.1.1. Type

The first work zone characteristics is the work zone type which is an indication of the expected level of impact a work zone will have on travelers. FHWA provides example definitions in the report *FHWA Work Zone Self Assessment Guide* [3]. *TAT Volume IX uses* the work zone types as defined in the Work Zone Self Assessment Guide which categorizes work zones as being one of four types based upon the magnitude of expected impact. The most severe work zone is a Type I work zone which have the most impact on the traveling public and usually occur over a long duration. Examples include the Springfield Interchange "Mixing Bowl" project in Northern Virginia and the Big Dig in Boston. A Type IV work zone will have little disruption on traffic flow as well as a short duration. Minor maintenance operations such as guardrail repair, sign repair, and mowing are typical Type I

Work zone type is often a strong indicator of the level of resources available to conduct a work zone analysis. The physical characteristics of the highway network may have a strong influence on the impacts caused by the project. For example, rural highways in sparsely populated areas may lack good alternate routes. In addition, the perceived impacts of a highway project may vary depending on local expectations about what is an acceptable level of congestion. For example, residents of a major metropolitan area may be more accustomed to driving in congested conditions than residents of a small town. Thus, selecting a transportation modeling approach for work zone analysis should not be based solely upon the work zone type but rather augmenting that decision with a consideration of the complete range of factors addressed in this document.

Type I—Type I work zones affect the traveling public at the metropolitan, regional, intra- and interstate levels over lengthy periods of time. They attract significant public interest, impact large numbers of road users, and are deployed at significant cost. The consideration of Type I work zones often includes quantitative analysis of work zone impacts for both internal agency use and shaping public expectation. In many cases, an impact assessment for these major projects is required by agency policy. The transportation modeling approach developed is tailored to the needs of the agency and resources available to perform the analysis. Analytical tool categories deployed to support Type I work zones include the full range of tools from sketch-planning and HCM methodologies to microscopic simulation. However, the tools are used quite differently depending on the decisions being supported. Type I projects typically cause both localized and area-wide impacts on travel demand and traffic. Often a Type I work zone will employ various work zone analysis tools including a regional travel demand model to estimate region-wide travel impacts, and then use a sketch planning tool to determine traffic impacts on individual highway segments. Given the complex nature of these work zone types, it is important that the agency identify its goals and objectives, needs, and available resources to identify the specific tool or tools that will provide the agency with the outcomes it desires. These projects have the largest potential impact and generally have more substantial resources allocated to the assessment of predicted work zone impacts. Examples of Type I work zones include:

- Woodrow Wilson Bridge in Maryland/Virginia/District of Columbia
- Central Artery/Tunnel in Boston, Massachusetts
- Springfield Interchange "Mixing Bowl" in Springfield, Virginia
- I-15 reconstruction in Salt Lake City, Utah.

WOODROW WILSON BRIDGE REPLACEMENT

The replacement of the Woodrow Wilson Bridge, along Interstate 95 in the Metropolitan Washington, DC area, is an extremely large and complex construction project requiring the close coordination of numerous contractors and various state and local government agencies. The Woodrow Wilson Bridge carries upwards of 100,000 vehicles per day along Interstate 95 which spans the Potomac River. The Woodrow Wilson Bridge Project spans a 7.5 mile-long corridor extending from the MD 210 interchange in Maryland to Telegraph Road in Virginia, crossing over the Potomac River. The current 6-lane bridge is being replaced with a dual-span bridge that will more than double the number of traffic lanes.

This project is categorized as the most severe work zone—Type I. This project directly impacted several miles of major roads and interchanges in two states including I-95/I-495 (Capital Beltway) from Maryland to Virginia, I-295, MD 414, and MD- 210 in Maryland, and Telegraph Road, Route 1, and Washington Street in Virginia.

To keep traffic moving throughout the multi-year project, the construction work for the Woodrow Wilson Bridge project had to be completed in individual phases instead of one large permanent work zone. Work zone plans had to be developed for each phase of this construction project.

The WWB project used a number of different modeling approaches to understand work zone impacts ranging from the simplest HCM methodologies to microscopic simulation primarily during the construction stage. More information can be found in the *Woodrow Wilson Bridge Reconstruction* case study.

Type II—Type II work zones impact travelers at the regional and metropolitan levels. They directly impact a range of road users and can attract significant public interest. The cost impacts are usually moderate to high and the work zones will usually be set up for a long period of time. Analysts considering Type II work zones will benefit from a quantitative analysis to evaluate impacts to the public. Often, this analysis is required by the agency. Similar to Type I, the full range of transportation modeling approaches are well suited for Type II work zones. This is not to suggest that any of the tools identified here are appropriate for a specific work zone project application. Rather, it is important to identify an appropriate role for a work zone analysis tool based upon the many opportunities and constraints available to the analyst. In some cases, the use of a specific transportation analysis tool will be required by agency policy. Examples of Type II work zones include:

- Major corridor reconstruction
- High-impact interchange improvements
- Full closures on high-volume facilities
- Major bridge repair
- Repaving projects that require long term lane closures

CALTRANS I-15 REHABILITATION

The existing four mile section of I-15 in Ontario, California consists of four to six lanes in each direction and carries an AADT of approximately 200,000, with a particularly high percentage of heavy truck traffic (about 12 percent on average on weekdays).

In addition, the corridor has consistently high weekday commuter peaks and similar volumes on weekends due to leisure travelers from Los Angeles headed to and from Las Vegas and resort locations along the Colorado River. The reconstruction will rebuild two to three truck lanes in each direction.

This construction project exhibits the characteristics of a Type II work zone because of the consistently high volumes and heavy truck traffic and the work zone impacts are moderate. Caltrans employed the use of three traffic modeling tools: a sketch-planning tool, an HCM model, and a mesoscopic traffic network analysis tool.
The sketch- planning analysis and HCM analysis were used to analyze alternative construction scenarios and the mesoscopic analysis tool was used to assist in the development of the transportation management plan and to provide supplemental construction staging analysis. More information can be found in the *Caltrans I-15 Pavement Reconstruction Project* case study.

Type III—Type III work zones have a low to moderate level of public interest. They can affect travelers at the metropolitan or regional level, but travelers are usually affected at a low to moderate level. The cost impacts will be less and the duration of the work zone will not be lengthy. The work zone may include lane closures, but these are implemented for only short periods of time. While the full spectrum of modeling approaches can be applied to Type III work zones, agency resources are frequently limited in nature when considering these types of projects.

Thus, the opportunity to set up and use more complex tools such as microscopic and mesoscopic simulation tools may be limited because of resource constraints.

A region may have a number of Type III work zones in the pipeline ready for planning, design, and construction. In this case, the use of simplified screening approaches (such as sketch planning and HCM tools) that can rapidly and cost-effectively analyze numerous different variations may be appropriate to determine those work zones requiring further and more detailed analysis. Examples of Type III work zones include:

- Repaving work on roadways and the National Highway System (NHS) with moderate Average Daily Traffic (ADT)
- Minor bridge repair
- Shoulder repair and construction
- Minor interchange repairs

ZION NATIONAL PARK

Zion National Park is an example of a Type III work zone. The repaving of the entrance booth area was of moderate public interest, primarily related to those businesses in the town of Springdale who would be impacted by any significant queuing caused by the temporary closure of entrance lanes. The National Park Service decided to use a sketch-planning tool during the final aspects of design to specifically address the scheduling of the work zone and determine whether or not it was justified to shift construction to night work. More information is available in the *Zion National Park: Entrance Booth Reconstruction* case study.

Type IV—Type IV work zones are expected to have little impact, if any, on the traveling public. They attract little public interest and work zone duration is short to moderate. It is sometimes difficult to provide information to travelers for these types of projects because they will usually be out of the roadway or off the shoulder before the message is relayed to the travelers. These types of work zones are usually mobile. While the full spectrum of transportation modeling approaches can be used to analyze a Type IV work zone, in all likelihood, many Type IV work zone projects may be effectively assessed either by a simple transportation modeling approach (e.g., sketch planning or HCM methodology) or the expert knowledge internalized by operations managers after years of experience. However, developing a strategic methodology to analyzing these types of work zones may be useful to ensure any roadwork does not unintentionally impact mobility. In this situation, it may be most cost-effective to establish an agency-wide program to assist personnel in quickly screening these types of work zones rapidly and effectively. Examples of Type IV work zones include:

- Certain low-impact striping work
- Guardrail repair
- Minor shoulder repair
- Pothole patching
- Very minor joint sealing
- Minor bridge painting
- Sign repair
- Mowing

NEW JERSEY TURNPIKE AUTHORITY: LANE CLOSURE APPLICATION

The New Jersey Turnpike Authority developed a GIS-based Lane Closure Application designed specifically to quantify the impacts associated with Type IV work zones, typically routine maintenance work that includes shoulder work or short-duration lane closures such as roadway sign or maintenance. A more detailed discussion of the tool is available in the *New Jersey Turnpike Authority: Lane Closure Application* case study.

2.1.2. Network Form

The second work zone factor that must be accounted for when deciding to use a work zone analysis tool is network form (also commonly referred to as network configuration or network structure). In a sense, network form is a surrogate for the overall complexity of the roadway where the work zone impacts will be evaluated and which are not accounted for within the FHWA work zone types. For work zone analysis, network form is composed of three general categories: isolated, pipe, and grid.

Isolated—An isolated network form consists of a single work zone that has limited interaction with the surrounding infrastructure and facilities. Because of the limited interaction, the work zone may only have periodic impact on the immediate surroundings depending on the size and duration of the construction project and the traffic demand for the facility. Examples of isolated networks include some bridge deck replacement projects, rural lane closures, or a redesigned interchange. Figure 6 provides a visual representation of two isolated work zones in the fictitious Hillsboro County. In the figure are two circles representing two work zones that are isolated from other roadway infrastructure as well as each other.

Pipe—A pipe network form (sometimes referred to as a corridor network form) consists of a roadway segment with two or more work zones interacting with each other and includes some limited access points and connections with other roadway segments (such as interstate entrance ramps). However, these work zones typically do not include any type of viable alternate or detour route. An example of a pipe network form would be a long interstate corridor reconstruction outside an urban area. Figure 7 provides a visual representation of a pipe network form work zone in the fictitious Hillsboro County. In the figure are three circles representing three work zones that are located close to each other resulting in an interaction among the three.

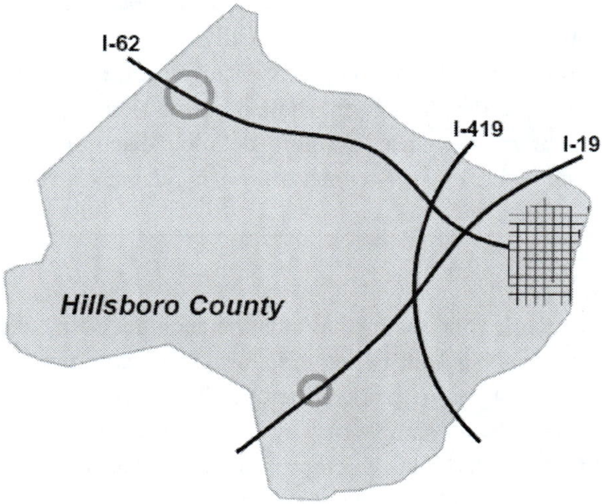

Figure 6. Isolated Network Form.

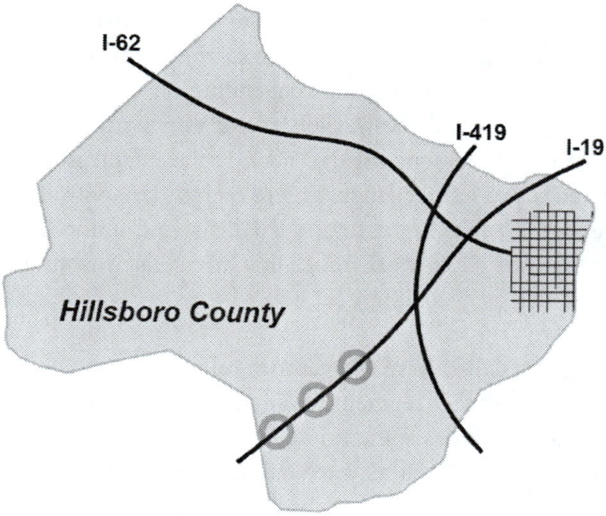

Figure 7. Pipe Network Form.

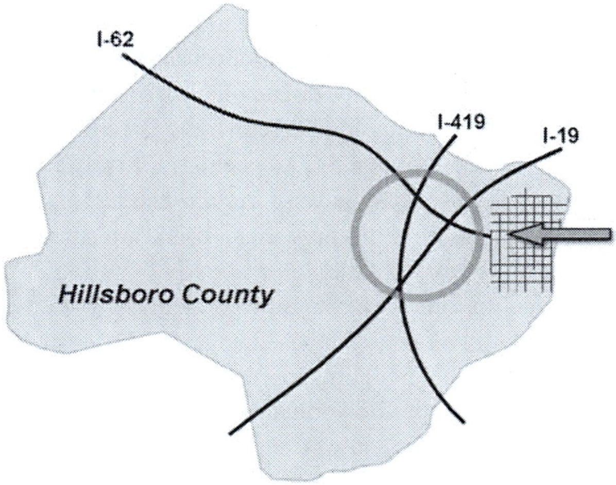

Figure 8. Grid Network Form.

Grid—A grid network form is a connected, inter-dependent structure with multiple access points and one or more viable alternate routes. Examples of grid network forms include urban interstate reconstruction, reconstruction involving full roadway closures, signalized arterial roadway reconstruction, and work zones located in urban centers. Figure 8 provides a visual representation of two grid network form work zones in the fictitious Hillsboro County. In the figure is one circle representing the location of a work zone involving three interstate roadways. Also indicated by the arrow is the downtown CBD.

The network form will have an impact on the effectiveness of various transportation modeling approaches. Work zones with impacts limited to a single mainline facility (e.g., isolated network form) are generally easier to model than more complex work zones with impact across a broader network of parallel facilities and cross-streets (e.g. grid network form). Many of the tools on the simpler end of the work zone modeling spectrum (e.g., HCM tools, or sketch-planning tools) assume that traffic demand follows a simple in/out pattern through the work zone. For isolated and pipe network forms, such an assumption is often reasonable. Sketch-planning tools have often been selected to analyze isolated work zones and there are numerous case study examples included in this document. A sketch-planning tool is often deemed a cost-effective approach in these cases, particularly if limited demand and capacity data are available.

The incorporation of detour and alternate routes (a major characteristic of grid network forms) limit which modeling approaches can be used. Many of the macroscopic simulation models do not have the ability to account for detour and alternate routes. Thus, if a detour route is required it is important the model selected has the functionality to account for traffic volume diverted to other facilities in the network because certain classes of vehicles are not permitted in the work zone or because road users are attempting to avoid delays from one or more work zones.

When work zone impacts create significant changes to an overall network traffic pattern (through diversion or other changes to routes normally taken in the network) a more detailed modeling approach must be considered. Often, an origin-destination pattern must be estimated. These potentially time-varying demand patterns are then input into a separate simulation model (typically microscopic or mesoscopic) to more realistically capture diversion effects. In very congested networks or networks with frequent fluctuations in demand or work zone capacity, the use of a dynamic traffic assignment (DTA) modeling component may be necessary. DTA can effectively account for the re-routing of traffic based upon rapidly changing conditions such as intermittent lane closures or congestion on detour routes. However, DTA functionality is only found on the more complex end of the work zone modeling spectrum with mesoscopic and microscopic models and adds considerably to the level of effort required for the study. More information on DTA can be found at http://ops.fhwa.dot.gov/ rafficanalysistools/dta.htm.

2.1.3. Geographic Scale

The third factor under work zone characteristics is the geographic scale of the work zone. Geographical scale refers to the size of the area impacted by the work zone. Simpler projects typically have a relatively small geographical scale. For example, a bridge replacement on a rural highway might affect only the road where the bridge is located and one or two nearby routes. At the other extreme, complex projects may influence the traffic patterns in a very large area. For example, the effects of a freeway-to-freeway interchange replacement project might extend all the way to the next freeway-to-freeway interchange.

Work Zone Size—The work zone size encompasses the physical construction zone and the immediate area near the work zone that will be affected by the implementation. These sizes refer to the physical size of the work zone and may or may not be related to the type of work zone. For example, a large construction project does not necessarily have to be a Type I work zone. The construction project may cover a large geographic area, but not impact a large number of travelers or be set up for a long duration. Work zones are divided into three sizes, small, medium and large.

- **Small**—A work zone implemented on a short segment of an individual roadway or a single intersection.
- **Medium**—One or more work zones on longer stretches of a single facility and/or portions of adjacent facilities.
- **Large**—Interacting work zones implemented on significant elements of a larger roadway network.

The size of the work zone will be directly related to the type of analysis that needs to be performed. In general, simpler tools can be used successfully for projects with small geographical scope, while more complex tools are necessary if the analysis extends over a large area. If the analysis area is quite large, meoscopic or microscopic traffic models are typically used. It is important to understand that as the geographical scale increases, so does the level of time, effort and skill required to calibrate the model to match real-world conditions. In resource- constrained cases, transportation modeling approaches may need to be limited to sketch-planning tools, HCM methodologies, or traffic signal optimization tools (if warranted). In other cases, a combination of tools can be used together to offset issues of scale and complexity. There is also the possibility of separating larger work zones into smaller ones, thus focusing resources on more critical elements of the overall work zone. In these situations, the analyst will have to be diligent about accounting for individual work zone impacts in a cumulative manner.

CALTRANS I-15 REHABILITATION PROJECT

An existing four mile section of I-15 in Ontario, California consists of four to six lanes in each direction and planned reconstruction will rebuild two to three truck lanes in each direction. This project exhibits the characteristics of a medium size work zone because it consists of a corridor that covers four miles of existing interstate and the work zone area encompasses three parallel routes to the east and west of the corridor.

This medium sized project was analyzed using sketch-planning, HCM, and mesoscopic simulation models.

> The mesoscopic model analyzed a variety of detour routes. More information can be found in the *Caltrans I-15 Pavement Reconstruction Project* case study.

Analysis Area Dimension—The analysis area dimension identifies the full area impacted by the work zone. In many cases, the impact of the project will extend beyond the limits of the work zone itself. This may seem self-evident to some analysts, but all too often it is overlooked when establishing the scope and budget for the work zone analysis, and selecting a methodology that matches the requirements of the project. For example, the construction project might encompass one or two large interchanges along an interstate corridor, but the impact of that project may extend to a larger area depending on traffic demand, lane and ramp closures, and detours and alternate routes. There are transportation modeling approaches that can evaluate each of these types of areas depending on the requirements of the agency. The analysis area dimension is divided into three categories: site, local, and metropolitan.

- **Site**—Restricted to the immediate area surrounding the work zone.
- **Local**—Includes surrounding area beyond the work zone to account for detour routes or other localized impacts.
- **Metropolitan**—Includes multiple jurisdictions (counties, cities, etc.) or facilities.

Table 1 provides a summary of the work zone characteristics as they relate to the project application case studies while Table 2 provides a summary of the work zone analysis approaches that have been historically used in previous work zone assessments. In most cases involving large or high impact projects, sketch planning tools should only be used to augment other tools as part of a multi-scale approach. For example, a valid approach would be to use a network-based model (such as an existing regional travel demand model) to determine the redistribution of traffic in the network, and then the HCM method or sketch planning tool would be used to analyze the localized impacts on affected links. Sketch planning tools cannot carry out traffic redistribution. Therefore, directly applying a sketch planning tool to a complex network carries significant technical risk: the tool is likely to over-estimate the congestion in the immediate vicinity of the work zone, and under-estimate the impacts of redistributed traffic farther away from the work zone.

MICHIGAN DOT: I-75 TRADE CORRIDOR MOTSIM

Michigan DOT is developing analytical capabilities to conduct a large-scale analysis of the 19-mile segment of I-75 between Detroit and the City of Pontiac.

The microscopic simulation network will consist of the entire metropolitan area associated with the 19- mile segment and builds upon previous analyses of adjacent roadways.

Currently in the planning stage, the project entails widening the road from three to four lanes and will enable Michigan DOT to test various scenarios to ensure that mobility is provided throughout the work zone, constructability is maintained, and staging is adequate.

More information can be found in the *Michigan DOT: Southeastern Michigan Simulation Network* case study.

Traffic signal timing tools have important applications for projects in urban and suburban areas. The signal optimization tool can be used to identify the need for temporary traffic signals and to revise the signal timing plans for existing signals where volumes are expected to increase or decrease as a result of the construction.

As with the sketch planning tools, a network-based analysis may be necessary to determine the new traffic pattern resulting from lane closures, link closures, or temporarily reduced link capacity. Once the new traffic pattern is developed with the network- based tool, specific timing plans can be developed to make the best use of the available capacity.

The suitability classification system has the following definition in Table 2:

● = Often analyzed by the transportation modeling approach.
⦸ = Sometimes analyzed by the transportation modeling approach.
○ = Rarely analyzed by the transportation modeling approach.
n/a = No examples of the transportation modeling approach being applied.

Table 1. Work Zone Characteristics Case Study Attributes

Case Study	Category	Modeling Approach	Project Life-Cycle Stage	Work Zone Characteristics			
				Type	*Configuration*	*Size*	*Analysis Area Dimensions*
CA I-15	Application	Sketch Planning Mesoscopic Simulation	Planning	II	Pipe	Medium	Metropolitan
GNP	Application	Sketch Planning	Planning	I	Pipe	Medium	Local
MD-LCAP	Approach	Sketch Planning	All	II, III, IV	Pipe	Small	Local
MDOT-SEMSN	Approach	Microscopic	All	I, II	Grid	Large	Metropolitan
MI AMB	Application	Microscopic Simulation	PE/Design	I	Grid	Medium	Metropolitan
MI I-75	Application	Microscopic Simulation	Planning	I	Grid	Large	Metropolitan
MI I-94	Application	Microscopic Simulation	Planning	II	Grid	Medium	Metropolitan
NJTA-LCA	Approach	HCM	Planning, Operations	IV	Pipe	Small	Local
NS-Reeves	Application	Sketch Planning	Planning	III	Isolated	Small	Local
UT I-15	Application	Travel Demand Model	PE/Design	I	Grid	Large	Metropolitan
Wisconsin	Application	Traffic Signal Optimization	Operations	III	Pipe	Medium	Local
WisDOT	Approach	Various	Planning	III, IV	Pipe	Small	Local

Case Study	**Category**	**Modeling Approach**	**Project Life-Cycle Stage**	**Work Zone Characteristics**			
				Type	***Configuration***	***Size***	***Analysis Area Dimensions***
WWB-LCA	Application	HCM	Construction	I	Pipe	Small	Site
WWB-RCA	Application	Sketch Planning	Construction	I	Grid	Large	Metropolitan
WWB-ROA	Application	Microscopic Simulation	PE/Design	I	Grid	Small	Local
YOS	Application	Sketch Planning	Planning	II	Pipe	Medium	Local
ZION	Application	Sketch Planning	PE/Design	III	Isolated	Small	Local

Table 2. Work Zone Characteristics Factors

***Factors* Work Zone Charateristics**	**Summary**	**Transportation Modeling Approach**						**Case Study**
		Sketch- Planning and HCM Methodologies	***Travel Demand Models***	***Traffic Signal Optimization***	***Macroscopic Simulation***	***Mesoscopic Simulation***	***Microscopic Simulation***	
Type								
Type I	Affects a large number of traveler at the regional scale, with high public interest, significant user cost impacts and a long duration.	●	⊘	⊘	⊘	●	●	GNP, MI AMB, MI I-75, UT I-15

Table 2. (Continued)

***Factors* Work Zone Charateristics**	**Summary**	**Transportation Modeling Approach**						**Case Study**
		Sketch- Planning and HCM Methodologies	***Travel Demand Models***	***Traffic Signal Optimization***	***Macroscopic Simulation***	***Mesoscopic Simulation***	***Microscopic Simulation***	
Type II	Similar to Type I but moderate impact.	⃠	⃠	⃠	⃠	●	●	CA I-15, MI I-94, YOS, MD-LCAP
Type III	Similar to Type I but low impact.	⃠	○	○	○	⃠	⃠	NS-Reeves, ZION, Wisconsin, MD-LCAP
Type IV	Short-duration work zone with low visibility and public interest.	●	○	○	○	○	○	NJTA-LCA, MD-LCAP
Network Configuration								
Isolated	Single work zone with limited interaction of surrounding infrastructure.	●	○	○	○	○	⃠	NS-Reeves, ZION, Wisconsin, MD-LCAP
Pipe	Roadway segment with multiple, interacting work zones.	●	○	○	⃠	⃠	●	CA I-15, GNP, WWB-LCA, YOS

Factors Work Zone Charateristics	Summary	Transportation Modeling Approach						Case Study
		Sketch- Planning and HCM Methodologies	***Travel Demand Models***	***Traffic Signal Optimization***	***Macroscopic Simulation***	***Mesoscopic Simulation***	***Microscopic Simulation***	
Grid	Connected, inter-dependent network structure with multiple access points and alternate routes.	○	⊘	⊘	⊘	●	⊘	MI AMB, MI I-75, MI I-94, UT I-15, WWB-RCA, WWB-ROA
Geographic Scale: Work Zone Size								
Small	A work zone implemented on a short segment of an individual roadway or a single intersection.	●	○	○	○	○	⊘	NS-Reeves, WWB-LCA, WWB- ROA, ZION, MD-LCAP
Medium	One or more work zones on longer stretches of a single facility and/or portions of adjacent facilities.	●	○	⊘	○	●	⊘	CA I-15, GNP, MI AMB, MI I-4,YOS, Wisconsin
Large	Interacting work zones implemented on significant elements of a larger roadway network.	⊘	⊘	○	⊘	⊘	○	MI I-75, UT I-15, WWB-RCA
Geographic Scale: Analysis Area Dimension								
Site	Analysis area is restricted to the immediate area surrounding the work zone.	●	○	○	○	○	⊘	WWB-LCA, Wisconsin, MD-LCAP

Table 2. (Continued)

Factors Work Zone Charateristics	Summary	Transportation Modeling Approach						Case Study
		Sketch- Planning and HCM Methodologies	*Travel Demand Models*	*Traffic Signal Optimization*	*Macroscopic Simulation*	*Mesoscopic Simulation*	*Microscopic Simulation*	
Local	Analysis area includes surrounding area beyond the work zone to account for detour routes or other localized impacts.	⊘	○	○	○	●	⊘	GNP, NS-Reeves, , WWB-ROA, YOS, ZION
Metropolitan	Analysis area includes multiple jurisdictions (counties, cities, etc.) or facilities.	⊘	⊘	⊘	⊘	⊘	○	CA I-15, MI AMB, MI I-75, MI I-94, UT I-15, WWB-RCA

●= Often analyzed by the transportation modeling approach.
⊘= Sometimes analyzed by the transportation modeling approach.
○= Rarely analyzed by the transportation modeling approach.
n/a = No examples of the transportation modeling approach being applied.

2.2. Transportation Management Plan Strategies[2]

Transportation Management Plan (TMP) strategies should be considered for all work zones, but the amount of effort and resources needed to produce the TMP will depend on the construction project and the characteristics of the work zone [4]. Smaller maintenance projects may utilize a simple TMP whereas a significant project that anticipates considerable traffic impacts requires a more detailed TMP. The use of a transportation modeling tool depends primarily upon the elements of the TMP. The following sections provide a discussion of the use of transportation modeling tools for the three categories of TMP strategies: Temporary Traffic Control, Public Information, and Transportation Operations.

2.2.1. Temporary Traffic Control

The TMP for all projects must contain a temporary traffic control (TTC) plan. TTC strategies include conceptual decisions on possible construction approaches, traffic control and management approaches, and time of construction. The scope of the TTC plan is dependent on the project characteristics and the traffic safety and control requirements required by the agency. The TTC plan consists of traffic control/design strategies, traffic control devices, and coordinating/ contracting/ innovation.

Control Strategies—Control strategies include traffic control approaches used to accommodate travelers within the work zone area, while providing adequate access to the roadway for the required construction work to be performed. The impacts of the work zone on traffic will be dependent on the types of control strategies that are established for a project. Control strategies for temporary traffic control include those listed below. Those in **bold** are control strategies that could be analyzed using the transportation modeling approaches described in Section 1.1:

- **Construction phasing/staging**
- **Full roadway closures**
- **Lane shifts or closures**
- **One-lane, two-way operation**
- **Two-way traffic on one side of a divided facility**
- **Reversible lanes**
- **Ramp closures/relocation**
- **Freeway-to-freeway interchange closures**

- **Night work**
- **Weekend work**
- **Work hour restrictions for peak travel**
- **Pedestrian/bicycle access improvements**
- **Business access improvements**
- **Off-site detours/use of alternate routes**

All of the control strategies can be analyzed using various modeling approaches. The decision to use a certain modeling approach is often based upon the stage in the project life-cycle. For example, the decision to use a certain construction technique requiring full roadway closure may rest upon whether or not the identified detour routes can handle the increased volume. In this case, the ability to use certain tools will be constrained by the decision to be made. Often, the use of a specific control strategy will be decided early in the decision-making process based upon a specific DOT policy either encouraging or disallowing a certain strategy.

Currently, the most prevalent modeling approaches used by analysts include sketch-planning/ HCM methodologies, mesoscopic simulation, and microscopic simulation tools. The tradeoff between these modeling approaches lies within the level of detail required and resources available. Microscopic and mesoscopic simulation tools can provide detailed analysis to evaluate detailed control strategy alternatives for the work zone. However, the resources and data requirements are significant. For example, an analyst may be required to determine allowable times for partial lane closures for a significant construction project during the PE/Design stage. Overall accuracy may be an important component to the overall cost of the construction project. In this case, a detailed modeling tool, such as microscopic simulation, may be warranted to provide the detailed analysis. However, time constraints may limit the analysis period to less than two weeks thus the use of a sketch-planning tool may be the best option. In the end, the modeling tool utilized may be determined in large part by the control strategy(ies) to be assessed and the ability of the modeling tool to provide useful results.

Michigan DOT: Ambassador Bridge Gateway MOTSIM

MDOT's Metro Region has a network simulation of an interstate freeway closure for the Ambassador Bridge Gateway Project (2008-2009).

Mobility plans were developed and implemented. A second model was run to incorporate other planned lane closures during the summer of 2008, and as expected the combined effects of the multiple closures undermined the original mobility plan. A third model for closures during the summer of 2009 is planned. However, in neither case were the 2008 and 2009 models ready in time to inform the project selection decisions. While these models are useful in allowing MDOT to tweak the TMP of the Gateway Project, they would have been much more useful if they had been performed in time to influence the project selection and the TMPs of the other projects. More information can be found in the *Michigan DOT: Southeastern Michigan Simulation Network* case study.

Traffic Control Devices—Traffic control devices used for temporary traffic control plans need to follow Part 6 of the Manual on Uniform Traffic Control Devices (MUTCD). The MUTCD provides standards and guidelines for utilizing traffic control devices within a work zone. In addition, other information is provided for installing, maintaining, and operating the work zone traffic control devices. The traffic control devices are applicable to all types of construction projects from small maintenance projects to significant construction projects. Traffic control devices utilized within work zones for temporary traffic control include those listed below. Those in **bold** are traffic control devices that have been analyzed using the transportation modeling approaches described in Section 1.1:

- Temporary signs
- **Changeable message signs**
- Arrow panels
- Channelizing devices
- Temporary pavement markings
- **Flaggers and uniformed traffic control officers**
- **Temporary traffic signals**
- Lighting devices

Three of the eight traffic control devices can be analyzed using various modeling approaches. Signing and marking cannot be analyzed with work zone analysis tools since the tools are based upon traffic flow theory. Thus, variables associated with the number of lanes open to traffic, lane width, free-flow design speed, percent trucks, etc. can be evaluated. In practical terms, the analyst can determine the impacts of using changeable message signs to convey traveler information, using flaggers for traffic control, and the use of temporary traffic signals:

- There are a handful of tools that can analyze the capacity of flagging operations.
- Traffic signal optimization tools are designed to analyze signal timing (permanent or temporary).
- There are many generic and specialized tools that can model reductions in the number of lanes and/or reduction in the capacity of the individual lanes.

Depending on other work zone characteristics, traffic control devices can be analyzed using the full spectrum of modeling approaches. Often, the selection of these devices will be made later in the project life-cycle, making it more likely that a previous modeling effort has taken place. Thus, an analyst may be able to leverage previous efforts. The ultimate decision to use one approach over the other will depend upon the resources available and the specific traffic control device to be included and whether the modeling approach is able to produce useful results.

GLACIER NATIONAL PARK

The Going to the Sun Road in Glacier National Park was slated for an extensive multi- year rehabilitation project over a seven to eight year period. Since the Going to the Sun Road is a key reason for visiting Glacier National Park and has no alternative route, the roadway must remain open throughout the project to both allow access to facilities and serve as a scenic roadway with minimal disruption and delay for park visitors.

The development of the TMP included temporary traffic signals. It is typical practice on the Going to the Sun Road to use a fixed timing plan for the control of short work zones and to vary the fixed timing plan over the course of the day for longer work zones.

In the Glacier case study, analysts investigated a complete range of control options for the temporary traffic signals from the simplest (single fixed plan) to the most complex (all actuated) and estimated how likely the 30 minute maximum user delay threshold might be exceeded over the life of the rehabilitation project.

Base signal timing in this case study represents a single fixed plan used in the night and weekend periods when flaggers are not present.

A sketch-planning analysis tool was used to identify projected delays and queue length over the course of the project and to assess the likely effects of actuated signal control for 2-way, 1-lane operations, as well as the impact of reduced travel demand. More information can be found in the *Glacier National Park: Going to the Sun Road Rehabilitation Project* case study.

Coordination/Contracting/Innovation—Project coordination strategies have the potential to reduce the mobility and safety impacts of the work zone. Contracting strategies involve agreements to reduce project duration and traffic impact. Innovative construction strategies include construction techniques that speed up the construction process to complete the project in a shorter period of time. Project coordination strategies include coordinating with other projects, utilities and right- of-way coordination, and coordination with other non-highway transportation facilities such as railroads and transit. Contracting strategies include design-build contracting, A+B bidding to reduce construction time, incentive/disincentive clauses, and lane rental—assessing a fee to the contractor for the time the lane is unavailable to traffic. Innovative construction techniques could include quick curing concrete or using precast items to minimize traffic restrictions and allow roadways to be reopened sooner.

Traffic impact analysis as it relates to coordination, contracting, and innovative construction strategies is more difficult to directly model using any of the transportation modeling approaches. If the agency has identified several projects that should be coordinated due to the potential of network-wide effects, then using a microscopic or mesoscopic simulation tool that can model all the projects would be a useful task to determine the traffic impacts overall. The analysis could possibly identify collaboration among projects that could be beneficial for the agency. Both contracting strategies and innovative construction strategies that result in shorter construction periods should be modeled to identify the most efficient approaches. There are some travel demand models and macroscopic models that might provide outcomes that would satisfy the agency's requirements to identify the impact of the work zone that may not be as resource intensive.

An important component to coordination, contracting and innovative construction strategies is the incorporation of road user costs. There are three main categories of road user costs: delay (extra travel time caused by the construction), vehicle operating costs (extra fuel consumed by detouring to a longer route) and safety (increased risk of crashes caused by detouring onto a road built to lower standards). By computing the road user costs associated with each alternative, it becomes straightforward to do a trade-off analysis. For example, one might find that building a project using the "standard" construction method has a direct cost to the agency of $1 million and creates $3 million of indirect costs to road users. However, the "quick" construction method has a direct cost of $1.2 million and reduces the impact on road users to $2 million. In that case, one could say that spending an extra $200,000 of the agency's money results in $1 million of savings to road users, a 5:1 benefit/cost ratio.

CALTRANS I-15 REHABILITATION PROJECT

The existing four mile section of I-15 in Ontario, California consists of four to six lanes in each direction and the reconstruction will rebuild two to three truck lanes in each direction. The project analysis was divided into two phases.

- Phase I used an innovative sketch-planning tool to evaluate a number of "what if" scenarios to identify solutions to balance schedule production, traffic inconvenience, and agency costs. This tool could also be easily integrated with other traffic simulation tools to quantify road user costs during construction.
- During Phase II, a mesoscopic simulation analysis was performed on the network to assist in the development of the TMP.

Caltrans used a variety of analysis tools to address different construction strategies and was able to use the results from these tools to help prepare an overall cost estimate analysis. More information can be found in the *Caltrans I-15 Pavement Reconstruction Project* case study.

2.2.2. Public Information

Many construction projects, especially those that will result in significant impact to the traveling public will benefit from including a public information component in the Transportation Management Plan. For significant projects, public information is a required component of a TMP. Providing specific information to the public that alerts them to potential impacts due to construction and associated work zones can help drivers make informed choices on when to travel to avoid construction or delays associated with construction or when to take alternate routes to avoid construction altogether. Keeping the public informed throughout the duration of the project will help an agency identify potential impacts and ensure effective mitigation strategies are implemented. Agencies should coordinate with their public information office in the early stages of project development. Public information strategies include both public awareness and motorist information strategies.

WOODROW WILSON BRIDGE REPLACEMENT

The replacement of the Woodrow Wilson Bridge (WWB), along Interstate 95 in the Metropolitan Washington, DC area, is an extremely large and complex construction project. The current 6-lane bridge is being replaced with a dual-span bridge that will more than double the number of traffic lanes. The WWB team relied upon an extensive public information campaign that included both public awareness strategies and motorist information strategies, to inform users of the bridge about planned, upcoming, and on-going construction activities

As part of their effort to inform users, WWB personnel employed the use of sketch- planning tools to estimate potential queuing and delay associated with several high- profile and extended weekend roadway closures. The results of the sketch planning tool (after initially being verified with smaller lane closure events) were disseminated to the public via various media sources and were designed to encourage travelers to avoid the area during the certain periods. More information can be found in the *Woodrow Wilson Bridge Reconstruction* case study.

Public Awareness Strategies—Public awareness strategies include different methods to educate and reach out to the public and the community pertaining to upcoming construction projects and related work zones. The agency should develop strategies that can continually be updated throughout the duration of the project. There are many strategies available to provide public awareness which include the following:

- Brochures and mailers
- Press releases and media alerts
- Paid advertisements

- Public information center
- Telephone hotline
- Planned lane closure Web site
- Project Web site
- Public meetings and hearings
- Community task forces
- Coordination with media/schools/businesses/emergency services
- Work zone education and safety campaigns
- Work zone safety highway signs
- Rideshare promotions
- Visual information (videos, slides, presentations) for meetings or for web-based dissemination.

None of these public awareness strategies have been directly assessed using a modeling approach. While some of these strategies (e.g., coordination with media, press releases) are low cost or even free, some of these strategies (e.g., paid advertising) can be costly to implement and are typically associated with major projects. More significant projects in large metropolitan areas will require an intense public awareness campaign, while projects with relatively low impacts will rely on less expensive public awareness strategies that utilize existing outreach mechanisms (e.g., DOT website) and other media. While no single tool can directly model these strategies, many modeling approaches are able to be used to conduct "what if" scenarios to determine what level of demand reductions are necessary to ensure mobility measures are met. A realistic estimate of the travel demand reduction (either through changing when a trip occurs, what mode is used or canceling a trip altogether) is an essential input for most work zone analyses tools. For example, in the Milwaukee area, recent experience indicates that major freeway reconstruction projects result in 30 to 40% reduction in peak hour demand. If we analyzed our Milwaukee freeway work zones without taking this into account, we would severely over-predict the impacts of the construction. In this situation, *ex-post* studies documenting demand reductions associated with completed projects using one or more of the strategies could be compared with the model outputs.

Motorist Information Strategies—Motorist information strategies focus on providing real time traveler information to travelers regarding the project work zone. Examples of motorist information strategies include those listed below. Those in **bold** are motorist information strategies that have been analyzed using the approaches described in Section 1.1:

- Traffic radio
- **Changeable message signs**
- Temporary motorist information signs
- **Dynamic speed message signs**
- **Highway advisory radio**
- Extinguishable signs
- Web-based highway information network
- **511 traveler information systems**
- Freight travel information
- Transportation Management Centers

Four of the 10 motorist information strategies have been analyzed using one of the modeling approaches. Generally, these analyses focused on the general use of traveler information (e.g., via en-route signs, cell phones, or pre-trip information from a web-site). Also, these analyses have been significant efforts using mesoscopic and microscopic simulation tools to model motorist information strategies such as dynamic message signs to determine the impact these strategies could have on traveler diversions either through or around work zone and surrounding area.

2.2.3. Transportation Operations

Transportation operations strategies include improved transportation operations and management of the transportation network to help mitigate work zone impacts. Transportation operation strategies include demand management, corridor/network management, work zone safety management strategies, and traffic/incident management and enforcement strategies. For significant projects, transportation operations is a required component of a TMP.

Demand Management Strategies—Demand management strategies use techniques that are intended to reduce the volume of traffic that travels through a work zone by encouraging the motorist to consider when and how they travel. Examples include diverting motorists to alternate modes of travel, shifting motorists to alternate routes, and encouraging off-peak travel trips. Demand management strategies include those listed below. Those in **bold** are demand management strategies that have been analyzed using the transportation modeling approaches described in Section 1.1:

- **Transit service improvements**
- Transit incentives
- Shuttle services
- Ridesharing/carpooling incentives
- Park and ride promotion
- **High occupancy vehicle lanes**
- **Tolling/congestion pricing**
- **Ramp metering**
- Parking supply management
- Variable work hours
- Telecommuting

Four of the 11 demand management strategies have been analyzed using one of the modeling approaches. Demand management strategies are usually reserved for large construction projects that have a lengthy duration because most of these strategies deal with altering the motorists travel behavior. However, it is often difficult to predict the impact that demand management strategies, such as transit service improvement, will have on roadway user behavior since changes to overall demand are estimated separately, with the analyst identifying likely impacts from a range of potential responses based upon other work zone performance assessments. Thus, the work zone analysis tends to provide the decision maker with a range of potential impacts based upon previous results.

Previous efforts to model these strategies have included the entire spectrum of modeling approaches except for traffic signal optimization. However, the level of detail provided by these various approaches varies significantly. Sketch-planning tools may be able to model the impact of improved transit service on overall traffic volume, but this is likely to be an external calculation (increased transit mode split) which is then entered into the model. The fidelity associated with travel demand models (due to their ability to analyze large geographic areas) limits their ability to provide detailed impacts of demand management strategies specific to the work zone site. Mesoscopic and microscopic simulation tools are better able to provide the level of detail necessary to account for the impact of all four demand management strategies but require a greater amount of detail and fine-tuning.

In the realm of transportation economics, models have been developed to estimate the impacts of all of the transportation policy changes listed above. STEAM (Surface Transportation Efficiency Analysis Model) and STM (Strategic Transport Model) are two examples. These models typically use cross-elasticities to estimate demand for certain policy instruments (e.g., mode shift to HOV, transit, or HOT lanes). For example, if carpooling incentives are implemented, the models estimate the resulting reductions in the number of solo drivers. Some travel demand forecasting packages also incorporate these techniques to some extent. The output from these economic models could be used to adjust the travel demand for the work zone analysis.

Corridor/Network Management Strategies—Corridor and network management strategies are used to optimize traffic flow through the work zone corridor and adjacent roadways using various traffic operations techniques and technologies. Corridor and network management strategies include those listed below. Those in **bold** are corridor/network management strategies that have been analyzed using the transportation modeling approaches described in Section 1.1:

- **Signal timing and coordination improvements**
- **Temporary traffic signals**
- **Street and intersection improvements**
- **Bus turnouts**
- **Turn restrictions**
- Parking restrictions
- Truck and heavy vehicle restrictions
- **Separate truck lanes**
- **Reversible lanes**
- Dynamic lane closure systems
- **Ramp metering/temporary suspension of ramp metering**
- **Ramp closures**
- Railroad crossing controls
- **Coordination with adjacent construction sites**

Ten of the fourteen corridor/network management strategies have been analyzed using one of the modeling approaches. Corridor and network management strategies can be very effective in improving traffic flow through and around a work zone. The full spectrum of modeling approaches can be used to model these strategies with microscopic simulation being a typical modeling approach that analysts employ when evaluating these strategies. Other tools, including sketch- planning, may be useful for better understanding the impact of adjacent construction projects.

Often, these network-based strategies and the use of a modeling approach will require that institutional barriers be broken down. For example, some agencies may already be using traffic signal optimization tools and can work with the planning division to develop alternate route optimization scenarios or to optimize the addition of temporary traffic signals in the system as part of a work zone. This will likely require coordination efforts between the traffic engineering, planning, and construction divisions within the agency and possibly even with some outside agencies (e.g., state construction project within a city jurisdiction for signal timing). Good working relationships and communication between the divisions and the agencies will improve efforts to optimize traffic flow through the work zone. These types of agency interactions will be discussed later in this chapter.

Work Zone Safety Management Strategies—Work zone safety management strategies address traffic safety concerns in work zones by analyzing devices, features, and management procedures. Work zone safety management strategies include those listed below. Those in **bold** are work zone safety management strategies that have been analyzed using the transportation modeling approaches described in Section 1.1:

- **Speed limit reduction and variable speed limits**
- **Temporary traffic signals**
- Temporary traffic barriers
- Moveable traffic barrier systems
- Crash-cushions
- Temporary rumble strips
- Intrusion alarms
- Warning lights
- Automated flagger assistance devices
- Project task force and committees
- Construction safety supervisor and inspectors
- Road safety audits
- TMP monitor and inspection team
- Team meetings
- Project on-site safety training

- Safety awards and incentives
- Windshield surveys

While work zone management strategies are considered significant strategies in the success of a construction project, the application of modeling approaches are infrequently applied to evaluate the impact of these strategies on the work zone area. Generally speaking, modeling tools are not directly applicable to evaluating the impact these strategies have on the work zone area. These strategies cannot be directly modeled to determine the impact the work zone will have on the surrounding area; however they are required to some degree on all construction projects. There are some specialized safety analysis tools available to analysts (e.g., QUADRO, Safety Analyst, SafeNet, and Road Safety Risk Manager) but these generally do not address the specific work zone safety strategies listed here.

The safety analysis module in Quadro is designed specifically for work zone applications. It contains an internal database of the average number of crashes per mile traveled for various facility types, with and without construction. This information is used to compute the expected number of crashes in the work zone (the probability that there will be a crash). Quadro also converts this to a monetary value by multiplying by the expected increase in crashes by the average comprehensive cost of a crash. Using this method, the analyst can prepare a broad comparison of the relative safety of two construction alternatives.

Traffic/Incident Management and Enforcement Strategies—Traffic and incident management and enforcement strategies include many techniques to manage the traffic operations of the work zone. Work zone traffic management strategies entail monitoring prevailing traffic conditions and making adjustments to traffic operations as necessary. Traffic incidents during construction are situations where specific management strategies are useful to minimize the exacerbation of other work zone impacts. Incident management and enforcement strategies involve improved detection, verification, response, and clearance of crashes, mechanical failures, and other incidents in work zones and on detour routes. Traffic and incident management and enforcement strategies include those listed below. Those in **bold** are traffic and incident management and enforcement strategies that have been analyzed using the transportation modeling approaches described in Section 1.1:

- **ITS for traffic monitoring and management**
- **Transportation management centers**
- Surveillance [Closed-Circuit Television (CCTV), loop detectors, lasers, probe vehicles]
- Helicopter for aerial surveillance
- Traffic screens
- Call boxes
- Mile-post markers
- **Safety service patrols**
- Total station units
- Photogrammetry
- Coordination with media
- **Local detour routes**
- Contract support for incident management
- Incident and emergency management coordinator
- Incident and emergency response plan
- Dedicated police enforcement – paid by project
- Cooperative police enforcement
- Automated enforcement
- Increased penalties for work zone violations

Traffic and incident management and enforcement strategies are significant to the success of a construction project; however, transportation analysis tools are infrequently used to assess the impact of these strategies on the work zone area. The full spectrum of modeling approaches (except traffic signal optimization models) may be applicable to evaluating traffic/incident management and enforcement strategies that the agency could consider; however, most of these strategies cannot be directly modeled to determine the impact the work zone will have on the surrounding area. Generally, the analysis is limited to conducting "what if" scenarios. One exception is the use of safety service patrols where both sketch-planning and more complex tools have been used to assess the impact of reduced clearance time on work zone performance measures such as delay.

Table 3 provides a summary of the transportation modeling approaches that have been historically used to evaluate various TMP strategies. The suitability classification system has the following definition for Table 3:

● = TMP components have often been evaluated by the transportation modeling approach.

⦸ = TMP components have sometimes been evaluated by the transportation modeling approach.

○ = TMP components have rarely been evaluated by the transportation modeling approach.

n/a = No examples of the transportation modeling approach being used to evaluate TMP components.

2.3. Data

One of the most important factors when deciding to use a transportation modeling approach to assess work zone impacts is the availability and quality of data regarding the project. Regardless, any work zone impact analysis will be a balance between the type of data that are needed, where the data come from, and the quality of the data. Often, an analyst will also have to account for the project development phase and, ultimately, the required precision of the results[3].

2.3.1. Type

Important elements to work zone analysis are data describing the roadway network, the traffic on the roadway network, and cost variables such as value of time and construction costs. The extent to which these data are needed depends upon both the stage of the project life-cycle and the modeling approach being used. The roadway characteristic data are generally easily available; the most difficult being the availability of signal timing plans. Travel demand characteristics are more difficult to obtain and there are no hard and fast rules regarding the use of this data when using a transportation modeling approach except that, in general, the more accurate the data the more accurate the results regardless of the modeling approach being used.

Roadway Characteristics—These data can be thought of as describing the roadway network infrastructure such as physical attributes (lane configuration, intersection location, traffic signal location, alternate routes, etc.) and policy attributes (estimated roadway capacity, signal timing plans, etc.). Together, these attributes are required in order to create the necessary network that the model will use to generate results. Most models and analysis tools will require some form of roadway characteristic data. The type and amount of data will vary by which tool is selected. Roadway characteristic data can usually be collected from construction drawings, field surveys, aerial photographs or geographic information system files. This data is usually sufficient for sketch-planning models, traffic signal optimization, and macroscopic and mesoscopic modeling. For microscopic modeling, to improve the accuracy of the information, the analyst may want to verify the roadway characteristics with field data.

Travel Demand Characteristics—These data describe how the roadway network infrastructure is utilized by the user. This would include traffic counts, vehicle demand data, and transit data if applicable. Travel demand characteristics pertain to vehicle and traffic demand data needed to assess the impact of the work zone. Traffic counts, vehicle mix, delay data, transit data, origin-destination data, etc. are the types of travel demand data that will be required to prepare an assessment model of the work zone. Like roadway characteristic data, all the models and analysis tools discussed in this report will need some form of travel demand characteristic data. The tools chosen will determine to what extent data is needed. Sketch-planning and HCM methodologies will require less data requirements, whereas macroscopic, mesoscopic, and microscopic simulation models may require more data. As stated previously, the more site-specific data provided, and the more accurate the data, the better the results of the model. If site-specific data is unavailable and the resources to collect travel demand data are not available, there are sources that can provide default values for some of these parameters. Many of the models and tools themselves will include default variables to be used in situations where data is not available. Using default data should be done cautiously, as the default values may not reasonably approximate the site-specific conditions.

Cost Attributes—These data include parameters such as the value of time for different vehicle classes (passenger cars and trucks), the cost for different construction alternatives, and the effects on revenue that local businesses might incur due to the work zone. There are a few analysis tools that use the sketch-planning or mesoscopic approach and can input different cost attributes to evaluate the value of time and cost of the impact the work zone will have on the surrounding environment. These attributes include the revenue from a local business that might be impacted by the construction zone, or the cost and time of vehicles or trucks that have to alter their schedules and routes because of the work zone, among other factors.

Table 3. TMP Strategy Factors

Factors TMP Strategies	Summary	Transportation Modeling Approach						Case Study
		Sketch- Planning and HCM Methodologies	***Travel Demand Models***	***Traffic Signal Optimization***	***Macroscopic Simulation***	***Mesoscopic Simulation***	***Microscopic Simulation***	
Temporary Traffic Control								
Control Strategies	Traffic control approaches used to accommodate road users within the work zone and adjoining corridor in an efficient and safe manner.	⊘	○	○	○	⊘	●	Michigan DOT: Ambassador Gateway Bridge MOTSIM Caltrans I-15 Pavement Reconstruction Project
Traffic Control Devices	Use of signs, panels, pavement markings, signals and lighting devices for temporary traffic control	⊘	○	●	n/a	⊘	⊘	Glacier National Park: Going to the Sun Road Rehabilitation Project
Coordination/ Contracting/ Innovative Construction Strategies	Project coordination, contracting strategies and construction techniques to improve work zone operations.	●	⊘	n/a	⊘	○	○	Caltrans I-15 Pavement Reconstruction Project Utah DOT I-15 Reconstruction Design-Build Evaluation

Table 3. (Continued)

Factors TMP Strategies	Summary	Transportation Modeling Approach						Case Study
		Sketch- Planning and HCM Methodologies	*Travel Demand Models*	*Traffic Signal Optimization*	*Macroscopic Simulation*	*Mesoscopic Simulation*	*Microscopic Simulation*	
Public Information								
Public Awareness Strategies	Methods to educate and reach out to the public, businesses and community concerning the road project and work zone.	⃠	○	n/a	n/a	n/a	n/a	Woodrow Wilson Bridge Replacement
Motorist Information Strategies	Provision of current/real-time information to road users regar ding the project work zone.	⃠	○	n/a	n/a	n/a	○	Woodrow Wilson Bridge Replacement
Transportation Operations								
Demand Management Strategies	Techniques intended to reduce the volume of traffic traveling through the work zone	⃠	●	n/a	○	●	●	Woodrow Wilson Bridge Replacement
Corridor/Network Management (traffic	One or more work zones on longer stretches of a single facility and/or portions of adjacent facilities.	⃠	○	n/a	○	●	●	Michigan DOT: Ambassador Gateway Bridge MOTSIM

Factors **TMP Strategies**	**Summary**	**Transportation Modeling Approach**						**Case Study**
		Sketch- Planning and HCM Methodologies	***Travel Demand Models***	***Traffic Signal Optimization***	***Macroscopic Simulation***	***Mesoscopic Simulation***	***Microscopic Simulation***	
operations) Strategies								
Work Zone Safety Management Strategies	Devices, features, and management procedures used to address traffic safety concerns in work zones.	⦸	n/a	n/a	n/a	n/a	○	Utah DOT I-15 Reconstruction Design-Build Evaluation Wisconsin DOT: Transportation Management Plan Development Process
Traffic/Incident Management and Enforcement Strategies	Strategies to manage work zone traffic operations.	⦸	○	n/a	○	⦸	⦸	

●= TMP components have often been evaluated by the Transportation Modeling Approach.

⦸= TMP components have sometimes been evaluated by the Transportation Modeling Approach.

○= TMP components have rarely been evaluated by the Transportation Modeling Approach.

n/a = No examples of the Transportation Modeling Approach being used to evaluate TMP components.

2.3.2. Source

There are many sources of data related to a work zone analysis. In general, these data sources are separated into two categories: primary and archived.

Primary—Primary sources of data include any number of data collection techniques whereby traffic data counts are collected in the field either using automated sensors such as video cameras, electronic toll collection (ETC) transponders, loop detectors, station counters, etc. Models that require field traffic counts for analysis are traffic signal optimization and mesoscopic and microscopic simulation. The more recent the data collection is performed, the more accurate the modeling results will be. Primary sources of data also include any number of analytical models available to an analyst from previous studies or other agencies such as a regional planning organization.

YOSEMITE NATIONAL PARK: YOSEMITE VILLAGE ROADWAY RECONSTRUCTION

The Federal Lands Highway Division used a combination of archived data and field traffic counts for the analysis of the Yosemite Village Roadway Reconstruction. FLH engineers conducted a first-cut distribution of hourly and daily travel demand factors obtained from a 1998 traffic study. These values were combined with two-short term collection activities to refine the hourly and daily distributions and establish a rate of travel demand growth from 1998-2004. The field traffic counts were conducted using portable traffic counters which are available agency-wide within FLH. More information can be found in the *Yosemite National Park: Yosemite Village Roadway Reconstruction* case study.

Archived—Archived data is used most often in sketch-planning and HCM transportation modeling approaches. Depending on the analysis tool chosen and the required accuracy of the results, it can also be used to some degree by travel demand models, traffic signal optimization models, and macroscopic, mesoscopic, and microscopic simulation models. There are many sources of archived data including public entities such as a regional planning organization or departments in a transportation agency as well as commercial vendors. Using archived data can be a cost effective solution to conducting a work zone analysis if limited resources are available or a quick decision needs to be made.

NEW JERSEY TURNPIKE AUTHORITY: LANE CLOSURE APPLICATION

Based upon HCM methodologies, the New Jersey Turnpike Authority Lane Closure Application relies on archived data in order to conduct analyses. Typically, these data are between one to two months old. More information is available in the *New Jersey Turnpike Authority: Lane Closure Application* case study.

2.3.3. Quality

An important component to using data is the quality of the data. The purpose of using data (be it primary or archived) is to prepare models that best represent actual field conditions resulting in an analysis producing the most accurate results possible. Primary data can often provide the most valuable data for modeling; however, collecting this data is resource intensive and costly. Many analyses for work zone planning purposes are conducted without this type of resource intensive primary data collection. In the case of significant Type I work zones that will affect a large number of travelers for long durations, field data collection may be beneficial to evaluate the work zone impacts more accurately. An analyst can benefit from visual observation because they can identify behavior in the field that is not as obvious in other data collection methods. This allows the analyst to adjust the model to better reflect existing conditions and improve the results when modeling for work zone implementation. In general, the analyst is concerned with the following four characteristics regarding the quality of the data: collection frequency, geographic coverage, archive length, and accuracy.

Collection Frequency—Collection frequency describes how often data is collected. This is important based upon the type of work zone being analyzed and the project stage. Work zone analyses occurring in the planning and design stage may need sample data collected only once in order to characterize daily travel demand or hourly roadway capacity to support a high-level decision about work zone scheduling. Projects in the construction stage may need data that is collected more frequently (e.g., traffic counts or travel times at five-minute intervals) to assess model outputs.

Geographic Coverage—An important consideration of data quality is the extent to which data is available for a specific location or region. Depending upon the location of the work zone, traffic data may not be available for a certain location (e.g., rural interchange or intersection). In these situations representative data may need to be used from other sources. Also, larger Type I work zones may need to have data available that encompasses an entire region rather than only surrounding the immediate vicinity of the work zone.

Archive Length—Archive length is concerned with the extent to which historical/archived data is available to an analyst. The length of the archive is important when temporal variations need to be considered as part of the work zone analysis. For example, work zones with a long durations occurring over multiple construction seasons will need to account for seasonal fluctuations in traffic demand as part of the analysis. Also, analysts considering the timing of a construction project in terms of taking place now or over the next five years will need to estimate the change in traffic demand over a number of years in order to project out future travel demand.

Accuracy—Accuracy is concerned with comparing outputs of the model with real-world conditions and the requirements of the analyst such that the model can be validated. In other words, what are the tolerances of the model? In some cases, such as a high-level analysis during the planning stage, a rough order of magnitude or directionality is sufficient. However, a work zone analysis occurring during the construction stage may require higher accuracies such that the analyst can estimate queue formation every 15 minutes. Accuracy will depend not just on the project stage but the measure of effectiveness as well. An analyst should always strive to identify and utilize accurate data. The key point with respect to understanding the inherent accuracy of a data source is to understand the level to which one can accurately apply the data. For example, a multi-year dependable archive of sample traffic counts taken a few times each year may be a useful and accurate source of data for estimating rough seasonal trends. However, these data may not be accurate enough to support a peak recurrent delay estimate for a specific day not in the sample set.

Woodrow Wilson Bridge Replacement

The WWB project was concerned with all four aspects of data quality as part of several of the modeling projects. For the microscopic simulation analysis, previous field observations were used to help calibrate the model.

In the case of using the sketch- planning tool for the first time, engineers ran the model to estimate queuing and delay and then verified those results in the field by observing the actual queue length and driving the work zone to measure travel times and determine delay. The initial field observations were used to ensure the model was calibrated correctly for future use. More information can be found in the *Woodrow Wilson Bridge Replacement* case study.

Table 4 provides a summary of the three categories of data factors associated with work zone modeling. ***Type*** provides an assessment of the level of detail historically used for each of the three data types: Roadway Characteristics, Travel Demand Characteristics and Cost Attributes. The suitability classification system has the following definition:

● = High level of detail historically used as part of the transportation modeling approach.
⦸ = Medium level of detail historically used as part of the transportation modeling approach.
○ = Low level of detail historically used as part of the transportation modeling approach.
n/a = Data type is not typically used as part of the transportation modeling approach.

Table 4, under the heading ***Source***, provides a summary of the sources of data commonly used for each of the transportation modeling approaches. The suitability classification system has the following definition:

● = Data source is commonly used for the transportation modeling approach
○= Data source is not commonly used for the transportation modeling approach.

Table 4, under the heading ***Quality***, provides an assessment as to the overall quality required for each of the quality aspects (collection frequency, geographic coverage, archive length, and accuracy). In all cases, it is important that the data available be of high quality such that the outputs of the model are useful in supporting the decisions being made.

Table 4. Data Factors

Factors Data	Summary	Transportation Modeling Approach: *Sketch-Planning and HCM Methodologies*	*Travel Demand Models*	*Traffic Signal Optimization*	*Macroscopic Simulation*	*Mesoscopic Simulation*	*Microscopic Simulation*	Case Study
Type	● = High level of detail historically used as part of the transportation modeling approach. ⊘ = Medium level of detail historically used as part of the transportation modeling approach. ○ = Low level of detail historically used as part of the transportation modeling approach. n/a = Data type is not typically used as part of the transportation modeling approach.							
Roadway Characteristics	Information regarding the physical design and operational characteristics of the roadway to be modeled.	○	⊘	●	●	●	●	New Jersey Turnpike Authority: Lane Closure Application
Travel Demand Characteristics	Information regarding vehicle and traffic demand associated with a roadway segment or system beginning analyzed.	⊘	⊘	●	●	●	●	Caltrans I-15 Pavement Reconstruction Project
Cost Attributes	Value of time for various vehicles classes (passenger cars and trucks) as well as revenue for local businesses impacted by roadway construction.	○	○	n/a	○	○	○	Yosemite National Park: Yosemite Village Roadway Reconstruction Woodrow Wilson Bridge Replacement
Source	● (grey) = Data source is commonly used for the transportation modeling approach. ○ = Data source is not commonly used for the transportation modeling approach.							
Primary	Roadway detection, probe vehicle, floating cars, ETC, analytical models (e.g., regional travel demand models), etc.	○	○	●	○	●	●	Woodrow Wilson Bridge Replacement
Archived	Using historic/archived traffic data available from planning agencies, transportation agency, or commercial vendors (e.g., traffic.com).	●	●	○	●	○	○	New Jersey Turnpike Authority: Lane Closure Application

Factors Data	Summary	Transportation Modeling Approach						Case Study
		Sketch-Planning and HCM Methodologies	*Travel Demand Models*	*Traffic Signal Optimization*	*Macroscopic Simulation*	*Mesoscopic Simulation*	*Microscopic Simulation*	
Quality	✓ = The transportation modeling approach requires high quality data. ✗ = The transportation modeling approach does not require high quality data.							
Collection Frequency	How often data is collected.	✓	✓	✓	✓	✓	✓	Woodrow Wilson Bridge Replacement
Geographic Coverage	Extent to which data is available for a specific region or location.	✓					✓	Woodrow Wilson Bridge Replacement
Archive Length	Extent to which historical/archived data is available.	✓					✓	Woodrow Wilson Bridge Replacement
Accuracy	Comparing model results with actual conditions to validate model	✓	✓	✓	✓	✓	✓	Woodrow Wilson Bridge Replacement

In all cases, it is important that the data available be of high quality such that the outputs of the model are useful in supporting the decisions being made

2.4. Agency Resources

Developing an analytical approach will be strongly dependent upon the resources the transportation agency has available and dedicates to these analyses. These resources can be categorized as institutional arrangements within the organization that enable or hinder the flow of information; technical staff that are able to conduct the required analysis and/or data collection; funding to acquire the technical expertise or models; and scheduling requirements. All four of these agency resource categories play an important role when making a determination to deploy a transportation modeling approach.

2.4.1. Institutional Arrangements

The Work Zone Safety and Mobility Rule is characterized by a policy-driven focus to institutionalize work zone processes and procedures, with specific provisions for application at the project-level [3]. This may mean involving additional agencies in the Planning, PE/Design, and Construction of roadway projects. Each agency needs to recognize that the traveling public is not concerned about lines on the map, but about moving quickly, safely, and efficiently through the network. Therefore, transportation officials need to focus on this system-level result.

For example, operational personnel at successful traffic management centers have found innovative ways to overcome institutional and technical limitations to maintain traffic incident management coordination. By developing small personal groups, a trust is established that permeates throughout the corresponding organizations. The co-location of agencies within an incident management center helps to strengthen the interagency relationships.

While there are many transportation and planning agencies which have developed comprehensive institutional arrangements regarding the use of travel demand modeling for short and long range planning, similar examples for work zone analysis are not all that common. Many of the Type I work zones (Mega Projects) have in place a mechanism to deploy the use of a transportation modeling approach as part of the overall project delivery process. Often, work zone analysis will be included. However, the smaller-scale projects such as Type II, III, and IV work zones do not necessarily include a budget for transportation modeling, let alone work zone analysis.

Often, the lack of an integrated or strategic methodology for work zone analysis can be characterized by four key institutional barriers:

- **Culture**—There exists within many transportation agencies a culture of compartmentalization whereby planners do not interact with designers who do not interact with operations personnel. In order for work zone impacts to be accounted for during the entire project life-cycle, these barriers need to be broken for a continuous flow of information and data.
- **Leadership**—Leaders are important to ensuring the successful deployment and use of transportation analysis tools for work zone analysis. Often, successful examples of work zone analyses will focus upon a single individual who took it upon themselves to change an organization's culture or institutional arrangements to enable the use of work zone analysis throughout the project life-cycle.
- **Data Management**—Because of the compartmentalization of many transportation agencies, it is sometimes difficult to acquire the necessary data. Often, an analyst will have to go to multiple departments or organizations in order to get the data (e.g., operations, planning, transit).
- **Contracting**—As discussed below, acquiring the technical expertise to conduct a work zone analysis may require the use of contractors with specialized knowledge and tools. Agencies will need to have in place flexible contracting arrangements (such as contractors on-call) whereby services can be obtained quickly and easily.

Working together to overcome institutional challenges and carefully planning and managing the challenges can provide significant benefits for all agencies involved. The benefits of establishing a strategic methodology for work zone analysis are much greater than what could be achieved through the deployment of a number of components operating in isolation. For example, deploying the use of a transportation modeling approach for work zone analysis early in the project life cycle (such as the Planning stage) can engage key players and stakeholders such that if future work zone analysis is needed later to answer questions on lane closures during the Construction stage, results can be generated quickly and cost effectively.

MICHIGAN DOT: SOUTHEASTERN MICHIGAN SIMULATION NETWORK

As part of developing the Southeastern Michigan Simulation Network (SEMSIN) Michigan DOT's Metro Region addressed many of the institutional barriers discussed above. As part of creating Regional Mobility Teams, the culture of the organization changed whereby planners, traffic engineers, and safety personnel now consistently work together to address mobility within work zones throughout the project life- cycle. Also, as part of creating the SEMSIN, a wealth of modeling data resources were discovered and are now readily available to any agency personnel for future use. The successful creation of the SEMSIN can be traced back to a single person who worked for its implementation and use. More information on the development of SEMSIN can be found in the *Michigan DOT: Southeastern Michigan Simulation Network* case study.

2.4.2. Technical Staff

A critical component to the successful use and deployment of work zone modeling tools is the technical expertise to use them. Often, transportation agencies will have either in-house expertise regarding the use of a specific model or access to a consultant with the expertise through a contractual arrangement. In this situation, an agency may be limited to a specific model based upon the technical capability at hand. For example, many transportation agencies make a decision to use modeling products from a specific vendor. In this case, if an agency/analyst wanted to use a different modeling tool that is not available from the vendor, additional time and resources would need to be made available in order to acquire and run the model. Even if an outside contractor is used to conduct the analysis, the transportation agency will still need to have enough expertise on- hand in order to oversee and validate the work zone analysis findings.

The other aspect to this is leveraging resources outside of one transportation agency. Many localities have a federally mandated regional planning organization (commonly referred to as a metropolitan planning organization or council of governments) chartered to conduct long-term planning analysis. In this situation, the planning organization may provide a base transportation network that could be used by the transportation agency in a work zone analysis.

2.4.3. Funding

Funding availability to conduct a work zone modeling analysis is another critical component of the agency resources. While there are many progressive transportation agencies trying to account for and mitigate the impact of work zones on the traveling public, an argument may still need to be made to allocate project funding to conduct a comprehensive work zone analysis. Sometimes, funding availability may limit the extent to which a work zone analysis can be conducted because money is not available for acquiring the necessary modeling tool or issuing a contract for the analysis to be conducted by a consultant. Regardless, any type of work zone analysis will require at least a minimum amount of funding and the accuracy and results of the analysis will depend to a certain degree on the amount of money available. As the use of more complicated tools is warranted, such as moving from a simplified delay estimation tool to a mesoscopic or microscopic transportation simulation tool, the costs will likely increase.

2.4.4. Schedule

A critical element in selecting a modeling approach is the project schedule. The amount of time required to conduct an analysis generally increases as one moves from the simplified to the more complex modeling approaches identified in Figure 1, the Work Zone Modeling Spectrum. In order to reduce the constraints of schedule to a work zone analysis, analysts should try to leverage other resources as much as possible. This includes seeking resources in-house that have conducted previous analyses related to the specific work zone area or general location as well as looking to other agencies and analyses that have used similar modeling tools. The point here is to leverage to the fullest extent possible the work that others have done so as not to "recreate the wheel".

Agencies who have taken a strategic approach to incorporating modeling into the overall work zone decision-making process have an advantage in addressing the schedule constraint. Agencies that have invested in developing modeling resources from the beginning (be it on-call contractors or in-house technical expertise) with a deployed capability in "stand-by" mode are better able address unanticipated questions when they arise and a decision must be made quickly. Those agencies that have invested in standing analytical capabilities will be more nimble in responding to these situations than those who start from scratch for each analysis. Thus, these agencies limit the impact that a schedule constraint has on generating useful results from the models as part of the decision-making process.

2.5. Work Zone Performance Measures

An important aspect in selecting a work zone analysis tool will be selecting a tool that outputs results that can be used to evaluate the work zone performance measures established by the transportation agency at the precision that is required to make a decision. Many transportation agencies use different types of measures of effectiveness for work zone impacts which can range from maximum additional travel time to maximum queue lengths caused by the work zone. Thus, if an agency has a requirement that no work zone may contribute more than 20 minutes of additional travel time delay, then the tool selected must be able to accurately predict that value. In addition, if the model will be used to assess financial incentives and disincentives, then it is imperative the model be able to predict the results accurately in order to set those levels.

2.5.1. Measures of Effectiveness

The following seven work zone measures of effectiveness come from the Work Zone Performance Assessment Measures compiled by FHWA in the publication *Work Zone Impacts Assessment: An Approach to Assess and Manage Work Zone Safety and Mobility Impacts of Road Projects (3)*. These seven were distilled from a larger list based upon the ability for currently available transportation analysis tools to calculate the performance measures.

Most of the transportation modeling approaches presented in Section 1.1 are capable of reporting the various work zone measures of effectiveness an agency requires. If agency policy stipulates a need to measure lane-by-lane queue lengths, the most accurate results will most likely be generated by a microscopic transportation simulation program since it provides results at the vehicle and lane level. However, microscopic tools are particularly complex and costly to deploy when modeling large corridors and regions. In these cases a mesoscopic model or regional planning model could be used to predict queue lengths when lane-level result are not critical. What is important is to select the approach that best provides the needed measures relative to the available data and work zone attributes.

Travel Time—A performance objective for many agencies might be to maintain an average travel time through the work zone for a certain period of time. Travel time is one of the most common measures used to track work zone delay. Microscopic and mesoscopic simulation can determine these results at the vehicle level, whereas sketch-planning, HCM methodologies, and travel demand models will provide travel time results based on the average for the system or work zone area. Some agencies may also collect real-time travel data during construction implementation and provide this information directly to travelers

UTAH DOT I-15 RECONSTRUCTION

See the *Utah DOT I-15 Reconstruction Design-Build Evaluation* case study for a discussion on the use of travel time as a performance measure.

Delay—To improve safety and mobility through a work zone, most agencies will measure delay through the work zone. Delay is defined as the amount of additional travel time drivers spend traveling which is caused by the work zone. Delay and travel time are directly related. Some agencies provide a delay rule that states a construction project cannot cause more than 20 minutes of additional delay or total delay through a work zone. Work zone analysis is used to estimate various delay measures (e.g., maximum or average delay) in order for construction to be staged accordingly. Mesoscopic and microscopic simulation models can model vehicle delay and total delay through the work zone. Sketch-planning tools, some travel demand tools, and some macroscopic simulation tools can model average vehicle or total delay through the work zone. Depending on the size and type of work zone and the data available for modeling, these tools may be sufficient for determining the impact the work zone has on delay.

YOSEMITE NATIONAL PARK

See the *Yosemite National Park: Yosemite Village Roadway Reconstruction* case study for a discussion on how a maximum user delay policy impacted the work zone.

Queue Length—Queue length is also a common performance measure that many agencies use to determine the impact of a work zone. Some agencies use queue length measures to determine when to set up lane closures or even when to remove lane closures if the queues get too long. Queue length is defined as the number of vehicles or length of vehicles queued to travel through the work zone due to the reduction of roadway capacity caused by the construction. Mesoscopic and microscopic simulation models can analyze queue lengths on a lane-by-lane basis. Data collection for these models will need to be more accurate so the model can properly represent the field conditions. Some sketch-planning tools and some macroscopic simulation models may also estimate queue length, representing the average queues for the time period. If detailed data is not available, sketch-planning models are often used for developing construction schedules that specify when or when not to close lanes.

ZION NATIONAL PARK

See the *Zion National Park: Entrance Booth Reconstruction* case study on the impact that queue length had in the decision to shift construction to night-time.

Speed—Speed is another traffic flow characteristic like delay that some agencies might measure to determine work zone impact. Similar to delay and travel time, speed is a common measure for assessing mobility through the work zone.

Results from modeling tools that determine speed can also be used to establish the speed limits for the work zone during design. There are many transportation modeling approaches to analyze speed. There are analysis tools within sketch-planning models, HCM models, travel demand models, macroscopic simulation, mesoscopic simulation, and microscopic simulation models that provide speed as an endogenous variable in the model output.

WOODROW WILSON BRIDGE RECONSTRUCTION

See the *Woodrow Wilson Bridge Reconstruction* case study for more information on the use of a microscopic simulation tool to evaluate variable speed limits within a work zone project.

Volume—Traffic volume as a performance measure is defined as the reduction in the amount of vehicles traveling on the roadway due to the work zone. Volume is directly related to capacity when determining the impact the work zone will have on the project.

Travel demand models, traffic signal optimization tools, and mesoscopic and microscopic simulation models all measure volume as it relates to the reduction of capacity through the work zone.

In all cases, volume is a count of vehicles passing over a certain point (e.g., vehicles per hour or vehicles per hour per lane). However, interpreting the volume value is different based upon where it is measured. For example, if the work zone to be analyzed creates a bottleneck on the roadway resulting in queuing, volume through the work zone is a measure of throughput. In situations where the work zone is not causing any queuing, volume is a measure of travel demand.

User Cost—User cost refers to the estimated loss of productivity from road user delay caused by the construction project and the work zone. User costs are more difficult to analyze and there are few analysis models that can provide this performance measure as a result. In addition some agencies find it difficult and more time consuming to place values on productivity loss. Road user costs might be considered early in the design stage to determine the best construction staging and they can also be used as an incentive or disincentive in contracts to reduce the overall construction schedule.

CALTRANS I-15 PAVEMENT RECONSTRUCTION PROJECT

See the *Caltrans I-15 Pavement Reconstruction Project* case study for a discussion on the use of cost data (including hard and soft costs) in selecting a work zone construction scenario.

Incidents—Incidents as a performance measure refers to the quantity of vehicle incidents that occur as a result of the construction project and the work zone area. When agencies establish goals and objectives to improve the safety and mobility through the work zone, the performance measures established will be related to traffic flow, delay and the number of incidents. There is always a strategic goal to reduce incidents in work zones; however there are very few models available to analyze how to reduce or mitigate incidents. Some agencies use historical data to analyze future work zones. There may be a microscopic simulation tool that can analyze this data depending on the needs of the agency. Determining the quantity of vehicle incidents will most likely be reserved for Type I or Type II construction projects.

UTAH DOT I-15 RECONSTRUCTION

See the *Utah DOT I-15 Reconstruction Design-Build Evaluation* case study for an example on how a crash analysis was used to influence roadway construction policy.

Cumulative Impact—The cumulative impact refers to assessing the impact of multiple or linking construction projects within a corridor or a region. This includes situations in a region where multiple construction projects occur simultaneously as well as a single construction project involving multiple work zones that may create an interaction with other. The impact globally may be more significant than the individual impact of each work zone and therefore needs to be considered. The Work Zone Rule definition of significant project addresses work zone impacts from either individual projects or the combined effects of a project's impacts with those of other concurrent projects nearby. Minimizing cumulative work zone impacts can be achieved by adjusting construction scheduling and project coordination. Mesoscopic simulation or sketch-planning models will provide the best results for evaluating the impact of multiple construction projects within a region. There may also be travel demand models or macroscopic simulation models that can perform this analysis depending on the needs of the agency.

MICHIGAN DOT

See the *Michigan DOT: Southeastern Michigan Simulation Network* case study for an example of addressing the cumulative impact of multiple construction projects in a region.

GLACIER NATIONAL PARK

See the *Glacier National Park: Going to the Sun Road Rehabilitation Project* case study for a detailed discussion on the impact multiple work zones have on work zone staging and design for a single project.

2.5.2. Precision of Results

The precision of results will be directly related to the modeling tool chosen to analyze work zone impacts. The size and characteristics of the work zone, the needs of the agency, the measures of effectiveness to be evaluated and those the model reports, and funding available for analysis will all determine which type of modeling analysis is best suited to examine work zone impacts. The stage or phase of the work zone may also affect which analysis tool is best for the situation. During the planning stage, the agency may be interested in high-level results, whereas during implementation of a large Type I construction project, more detailed results may be required when more data is available.

While the need for precision is often driven by the stage of the project life-cycle (Planning, PE/ Design, or Construction), other external factors related to agency resources (discussed in Section 2.4) may limit the available transportation modeling approaches, thus the overall level of precision. A summary of these tradeoffs is presented in Figure 9 below.

The bottom bar shows the level of precision generally needed for the project life cycle stages with more precision desired as one moves from Planning to PE/Design to Construction. The top bar shows a progressive refinement of impact parameters (data quality and data availability) to enable more and more precise work zone impact estimation.

Projects within the domain of the Planning stage generally need less precise results and, therefore, less precise data. When rough estimates are required, it is often worthwhile to consider existing data sources when getting started with an analysis. In this case the cost to collect detailed data is often unwarranted and existing data assimilated from multiple sources, such as template libraries and historical data, are generally sufficient to conduct an analysis to support rough estimation.

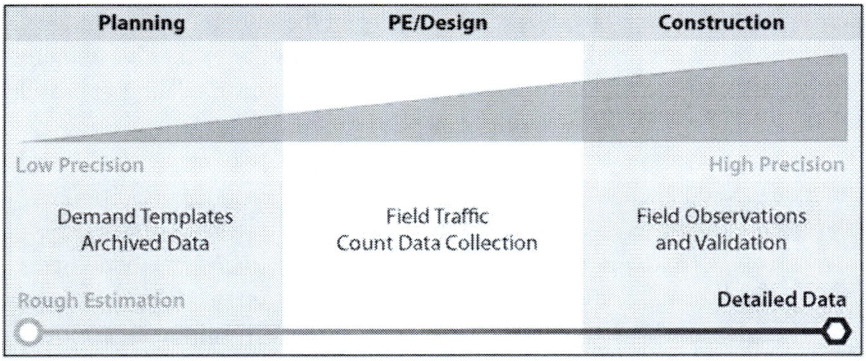

Figure 9. Work Zone Data Spectrum.

Projects in the PE/Design stage would require more refined and accurate data than those in the planning stage but not as detailed as those within the construction stage. In this stage of development, depending on the situation, historical data combined with more recent traffic count or visitor data may be sufficient to get the desired results. In some cases it is warranted to use some of the available resources to collect more detailed field data, especially if performance- based contracting is to be used or if liquidated damages will be established, in which case the most detailed data available needs to be used. On the other hand, projects entering or within the Construction stage would need much more accurate and timely data since the analysis will more than likely require differentiation between very similar alternatives that require a higher fidelity of data. For example, a contractor may request an additional lane of traffic be taken away for an extra two hours each Sunday morning in order to more quickly perform the work. Analyzing this type of request is a situation where highly precise results would be warranted in an operational work zone. In this situation, it would be possible to collect field data such as first-hand queue observations and traffic volume. Overall, the amount and types of data required will depend primarily upon the characteristics of the work zone, the analysis area dimensions, the needs and resources of the agency, and the accuracy required to evaluate the performance measures. Data collection will involve a combination of existing and available sources, and new data collected before and during work zone implementation.

An example of the precision of results is the difference between outputs from a sketch-planning tool and a microscopic simulation model. Sketch-planning tools designed specifically for work zone analysis (such as QUEWZ-98 or QuickZone) often provide results in one hour time slices. However, microscopic simulation models provide results on the minute-by-minute level. Thus, if a particular work zone analysis requires results every five minutes (such as a variable speed limit system), a transportation modeling approach providing higher precision of results may be required.

Modeling analysis results can be divided into two categories: rough estimation and detailed assessment.

- ***Rough Estimation***—Sketch-planning tools, HCM methodologies, travel demand models, macroscopic simulation models, and some traffic signal optimization tools will produce results that are often expressed as "maximum" or "average". These tools do not analyze traffic at the individual vehicle level. Results are considered average for the time period and the network and give the analyst an idea of what traffic impact might be like during work zone implementation. These tools are usually less data intensive and may provide enough detail for many work zone analyses.

- ***Detailed Assessment***—Mesoscopic and microscopic simulation models and some traffic signal optimization tools provide detailed results. Tools in these categories provide results that model individual vehicle movements and turning patterns. These tools are usually more resource intensive and require much more data. The results provided give a detailed assessment of existing conditions and multiple work zone alternatives and provide the agency with a good representation of what field conditions will be like during work zone implementation. Detailed assessments are often reserved for large, Type I work zones that will affect large numbers of travelers for a long duration. Another possibility for using mesoscopic or microscopic simulation would be if the agency already has a simulation model prepared for the region or area where the construction will be performed. The analyst can populate the model with the alternatives and new data to determine the best work zone alternative.

Table 5 provides a summary of the transportation modeling approaches that have been historically used to calculate the measures of effectiveness as well as the precision of results generated by the modeling approaches. The suitability classification system is different for each. The measures of effectiveness use the following definition in Table 5:

● = Performance measure is often calculated by the transportation modeling approach.

⦸ = Performance measure is sometimes calculated by the transportation modeling approach.

○ = Performance measure is rarely calculated by the transportation modeling approach.

n/a = Transportation modeling approach should not be used to calculate the performance measure.

Table 5. Performance Measure Factors

Factors Performance Measures	Summary	Transportation Modeling Approach						Case Study
		Sketch- Planning and HCM Methodologies	*Travel Demand Models*	*Traffic Signal Optimization*	*Macroscopic Simulation*	*Mesoscopic Simulation*	*Microscopic Simulation*	
Measure of Effectiveness. ●= Performance measure is often calculated by the transportation modeling approach. ⊘= Performance measure is sometimes calculated by the transportation modeling approach. ○= Performance measure is rarely calculated by the transportation modeling approach. n/a = Transportation modeling approach should not be used to calculate the performance measure.								
Travel Time	Amount of time it takes vehicles to travel through the work zone.	●	⊘	n/a	○	●	●	Utah DOT I-15 Reconstruction Design-Build Evaluation Glacier National Park: Going to the Sun Road Rehabilitation Project Caltrans I-15 Pavement Reconstruction Project
Delay	Amount of time drivers spend in traveling through the work zone.	●	⊘	n/a	○	●	●	Yosemite Nation Park: Yosemite Village Roadway Reconstruction

Factors **Performance Measures**	**Summary**	**Transportation Modeling Approach**						**Case Study**
		Sketch- Planning and HCM Methodologies	***Travel Demand Models***	***Traffic Signal Optimization***	***Macroscopic Simulation***	***Mesoscopic Simulation***	***Microscopic Simulation***	
Queue Length	Number of vehicles or total length of vehicles queued to travel through the work zone due to reduction in roadway capacity caused by the construction.	●	n/a	n/a	○	●	●	Zion National Park: Entrance Booth Reconstruction Woodrow Wilson Bridge: Roadway Closure Analysis
Speed	Vehicle speed through the work zone.	⊘	○	n/a	○	●	●	Woodrow Wilson Bridge Replacement
Volume	Reduction in amount of vehicles operating on the roadway due to the work zone.	⊘	○	n/a	○	●	●	
User Cost	Estimated productivity loss due to driver delay caused by the work zone.	⊘	○	n/a	○	○	○	Caltrans I-15 Pavement Reconstruction Project
Incidents	Quantity of vehicle incidents that occur as a result of work zone project.	○	○	n/a	○	○	○	Utah DOT I-15 Reconstruction Design-Build Evaluation
Cumulative Impact	Assessing the impact of multiple roadway construction	●	⊘	n/a	○	●	○	Glacier National Park: Going to the Sun Road

Table 5. (Continued)

Factors **Performance Measures**	**Summary**	**Transportation Modeling Approach**						**Case Study**
		Sketch- Planning and HCM Methodologies	***Travel Demand Models***	***Traffic Signal Optimization***	***Macroscopic Simulation***	***Mesoscopic Simulation***	***Microscopic Simulation***	
	projects within a region or corridor.							Rehabilitation Project Michigan DOT: I-94 Rehab MOTSIM
Precision of Results. ✓= The transportation modeling approach *commonly produces* results at this level of precision. ✗= The transportation modeling approach *does not commonly produce* results at this level of precision. Approach.								
Rough Estimation	Results produced by traffic analysis tools that do not model individual vehicles or turning movements. Results often expressed as "maximum" or "average".	✓	✓	✓	✓	✓	✗	New Jersey Turnpike Authority: Lane Closure Application Zion National Park: Entrance Booth Reconstruction
Detailed Assessment	Results produced by traffic analysis tools modeling individual vehicle movements and turning patterns.	✗	✗	✓	✗	✓	✓	Michigan DOT: Southeastern Michigan Simulation Network

The precision of results use the following definition in Table 5:

✓ = The transportation modeling approach ***commonly produces*** results at this level of precision.

✗ = The transportation modeling approach ***does not commonly produce*** results at this level of precision

3.0. Establishing a Strategic Methodology for Work Zone Analysis

Section 2.0 provided a detailed discussion of the myriad factors that influence which type of transportation modeling approach to choose. As observed earlier, the need to conduct a work zone analysis can occur at any stage during the project life cycle including Planning, PE/Design, and Construction. This breadth of application is evident in Table 1 which lists each of the case studies and where during the project life-cycle the analysis took place. Of the 13 project application case studies, seven occurred during Planning, four during PE/Design, and two during Construction. Clearly, the need for work zone analyses is not limited to a single stage. Thus it becomes important to develop a strategic methodology to applying transportation modeling approaches throughout the work zone project life-cycle.

To that end, this section deals with establishing a strategic methodology for work zone analysis in order to maintain analytical consistency when work zone planning and analysis spans the project life-cycle. As mentioned previously, the benefits of establishing a strategic methodology for work zone analysis are much greater than what could be achieved through the deployment of a number of components operating in isolation. Thus, strengths and weaknesses of strategic methodologies that utilize more generalized transportation analysis tools (e.g., sketch-planning and HCM) to screen projects for issues and then commit to more detailed transportation analysis tools are contrasted against other strategic methodologies where one (or more) transportation analysis tools are initially developed in detail in the Planning stage and utilized consistently as the project evolves. Three general strategic methodologies are presented in Figure 10 and described below:

- **Mono-Scale**—The *mono-scale* methodology features a single transportation analysis tool applied consistently throughout the project life-cycle. This is represented by the single bar in the figure. *Michigan DOT's SEMSIN* is an example of a Mono-Scale methodology.

- **Screening**—The *screening* methodology utilizes a series of transportation analysis tools throughout the project life-cycle, in an attempt to match the best transportation analysis tool to the specific decision being supported. This is portrayed by the vertically segmented bar in the figure with each segment representing a different transportation modeling approach being used based upon the project life-cycle stage. The *Wisconsin DOT: Transportation Management Plan Development Process* is an example of a Screening methodology.
- **Multi-Scale**—The *multi-scale* methodology involves the deployment of multiple transportation modeling approaches in an integrated and strategic way to support decision making throughout the project life-cycle. This is portrayed by the horizontally segmented bar in the figure with each segmented representing a modeling approach to be used based upon the need rather than the project life-cycle stage. The *Woodrow Wilson Bridge Reconstruction* is an example of a Multi-Scale methodology.

Just as no single transportation analysis tool is right for all work zone analyses, no single work zone analysis methodology represents the single best match for all projects. Sections 3.1 through 3.3 provides discussion about the strengths and weaknesses of these various strategies relative to the analysis factors (primarily the agency resources and performance measures) discussed in Section 2.0.

3.1. Mono-Scale

The mono-scale methodology features a single transportation analysis tool applied consistently throughout the project life-cycle. The advantage of this methodology is that model consistency is essentially guaranteed throughout the project life-cycle. Resources invested in developing the tool and its supporting data are passed along as the project proceeds. The disadvantage of this methodology is that a single tool may be well-suited to support some types of decisions but not well-suited to others. For example, a mono-scale methodology using microscopic simulation of the immediate work zone area may capture detailed elements of traffic control (work zone geometry and traffic signal operations) but have only limited capability to predict impacts from diversion on parallel routes. A mono-scale methodology may be an attractive option if a calibrated, well-maintained model of the work zone impact area is already in hand. If there is not a good match between the transportation modeling approach and model type in-hand, and there are some resources in hand to extend or enhance the analysis, a multi-scale methodology can be considered.

MICHIGAN DOT: SOUTHEASTERN MICHIGAN SIMULATION NETWORK

Development of Michigan DOT's SEMSIM Network was undertaken, in part, to leverage the up-front costs associated with developing a microscopic simulation network across multiple projects *as well as* multiple stages within the project life- cycle. The Michigan DOT Detroit Metro Region expended lots of resources (both time and money) in developing the initial application of the SEMSIM Network for the Ambassador Gateway Bridge MOTSIM during the PE/Design stage and will continue to use the network as needed during construction. The same network was expanded to use for two other projects, I-75 Trade Corridor and I-94 Rehab, which are in the Planning and PE/Design stages, respectively. More information can be found in the *Michigan DOT: Southeastern Michigan Simulation Network* case study.

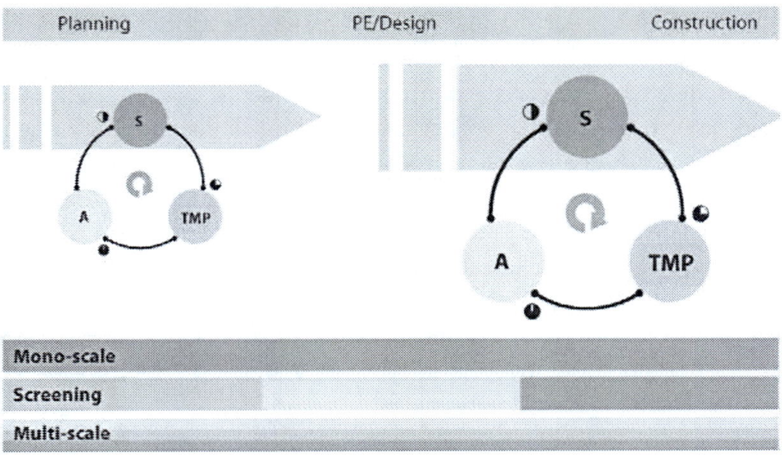

Figure 10. Comprehensive Work Zone Analysis Methodologies.

3.2. Screening

This methodology is characterized by the use of simpler tools using archived data to screen work zone projects to determine which may warrant more in-depth analysis. This concept has been used to allocate resources for work zone analysis when looking across a number of projects to identify potential mobility impacts. For single projects this includes the use of simpler transportation modeling approaches such as sketch-planning and HCM in early project phases and the use of a more complex transportation modeling approaches (if warranted) later in the project.

The advantages of this methodology include the focusing of agency resources (institutional, financial, and technical) on projects or issues that have the most potential mobility impact. It can be difficult early in project development or when looking across many projects to accurately identify where significant congestion is likely to be induced by work zone activity. The screening method allows for a uniform method to find the locations, times, and projected impacts as early as possible so these issues can be dealt with before the work zone decision engine has generated significant momentum.

WISCONSIN DOT: TRANSPORTATION MANAGEMENT PLAN DEVELOPMENT PROCESS

The Wisconsin DOT has established a screening approach in order to provide traffic engineers and planners in the various regional offices guidance on the selection of an appropriate transportation modeling approach. The screening approach developed by Wisconsin DOT is used to filter through those projects requiring more in-depth analysis (what they classify as Type IV and III work zones). If a project does require in-depth analysis (typically the use of microcopic or mesoscopic simulation tools) then a secondary model scoping worksheet is employed to further determine data needs and requirements.

To further reinforce the use of overall work zone analysis and the application of transportation models *throughout* the project life-cycle, Wisconsin DOT developed a training course which includes an exercise in applying transportation analysis models for a simple bridge replacement project beginning in the planning stage.

More information can be found in the *Wisconsin DOT: Transportation Management Plan Development Process* case study.

The disadvantage of this methodology is that the high-level screening transportation analysis tools used in earlier stages may not produce consistent predictions of impacts with more detailed tools used in later stages. The detailed analysis in later stages may reveal a nuance to the problem that may be contradictory to the direction the analysis with the screening tool has directed the decision engine. For example, using archived data in the planning stage, a screening tool may indicate that mid-day lane closures will not result in congestion. Thus, scheduling and application decisions later made in the PE/Design stage may count on a 4-6 hour window for mid-day work zone activity. However, a more detailed analysis with field traffic counts (e.g., travel demand data) collected specifically for the project in the PE/Design stage could indicate that unacceptably long delays are likely to result from mid-day lane closures.

3.3. Multi-Scale

This methodology deploys several tools in combination from project planning through construction, and any inconsistencies between models are identified and resolved throughout the process. This type of methodology is effective for looking at a complete range of issues using the most appropriate tool with the added value of cross-validation among tools and data types. The disadvantage of this methodology is that the agency resources (the technical staff, funding, and scheduling) are significant as well as the required data (specifically its quality and availability). This type of methodology is likely to be deployed only on the largest projects with a large work zone impact area, longer duration, and significant resources devoted to mobility impacts mitigation.

WOODROW WILSON BRIDGE REPLACEMENT

Planning for the reconstruction of the Woodrow Wilson Bridge began in the 1990s with record of decision on the locally preferred alternative made in November 1997. Construction began in October 2000 with final design completed in June 2003.

Construction of the bridge is planned to be completed in December 2009. All throughout this time, transportation models were being used for both the design of the facility itself as well as the design of the work zones to be implemented as part of the construction. The use of transportation models for work zone analysis primarily occurred during the construction stage as refinements to the staging of the various work zones were required. During this time, the transportation consultants used a variety of modeling approaches depending upon the requirements of the TMP, available data, and required performance measures. More information can be found in the *Woodrow Wilson Bridge Replacement* case study.

4.0. Identifying a Transportation Modeling Approach for Work Zone Analysis

The preceding two sections provided a detailed overview of the factors that influence the development of a transportation modeling approach and three strategic methodologies applied to maximize value from work zone analyses. Once an analyst has a better understanding of the factors influencing an overall work zone analysis, the next step is to consider two critical questions:

1) *Does it make sense to move forward with work zone analysis for a given project?*
2) *What criteria should be used to select an actual modeling approach for a project?*

The following two sub-sections discuss each of these questions in more detail. Section 4.1, Deciding to Analyze or Not Analyze, provides a framework to better understand the technical risk and rewards associated with ultimately deciding to deploy a transportation modeling approach. Section 4.2, Characterizing Case Studies by Transportation Modeling Approach, uses the same framework to classify the transportation modeling approaches utilized in the selected work zone analysis case studies. Finally, Section 4.3, Developing a Transportation Modeling Approach, provides a set of worksheets to guide an analyst in developing an appropriate transportation modeling approach and strategic methodology.

4.1. Deciding to Analyze or Not Analyze

In order to better understand which work zones warrant analysis, and to what extent, an analytical work zone decision framework was created. Figure 11 provides a notional representation of this framework for the work zone analyst considering the complex tradeoff between analytical opportunity and technical risk. Managing these two concepts is based on two high-level considerations shown on the two axes of Figure 11 the level of detail required to support decision- making on the X-axis (Level of Detail), and resources available to conduct a work zone analysis on the Y-axis (Resource Availability). The axes were developed based upon the five categories of work zone analysis factors discussed in Section 2.0 whereby resource availability incorporates data and agency resources, and level of detail incorporates work zone characteristics, desired TMP strategies, and required work zone performance measures. Effectively managing this tradeoff between analytical opportunity and technical risk is one of the most important responsibilities of the work zone analyst. Failure to realistically account for technical risk is the most frequent underlying issue when analytical efforts fail to meet the expectations of both analysts and decision-makers. This failure can manifest itself in failure to meet schedule, budget overrun, or an abandonment of the technical approach even after the expenditure of significant agency resources.

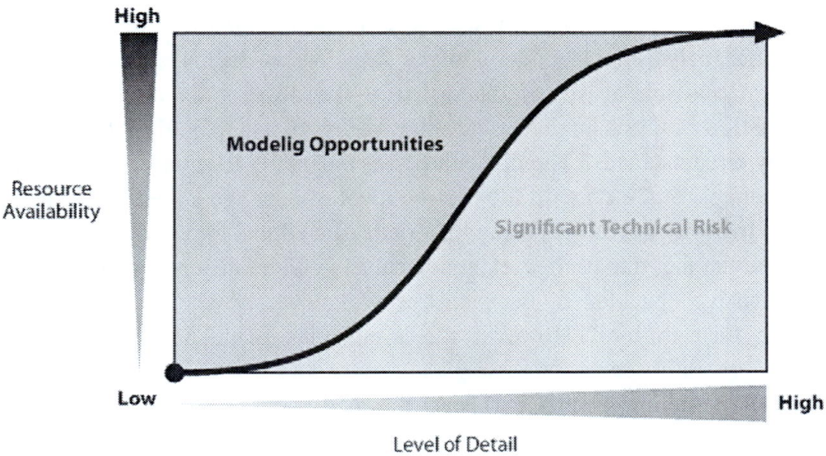

Figure 11. Analytical Work Zone Decision Framework.

As seen in Figure 11, when significant data and staff resources are available, there are many more options in matching a transportation modeling approach with the decisions to be supported. When resources are limited, however, the analyst must consider whether a less detailed analysis using a simpler tool is adequate to the requirements of the analysis. If the mis-match between the necessary level of detail and resource availability is high enough, the corresponding technical risk may be so high that a drastic re-consideration of modeling approach may be appropriate.

In some cases, the technical risk of attempting to conduct a work zone analysis with little reliable data or too few resources may outweigh the benefits of conducting it. In these cases, there is a risk that rushed or poorly-informed work zone analyses will provide highly inaccurate or misleading assessments of mobility impacts. Such analyses are to be avoided, since they not only represent a waste of project resources, but also because they may entrench institutional mistrust of transportation analysis tools and analytical results, even in cases where well-calibrated models do provide meaningful results.

4.2. Characterizing Case Studies by Transportation Modeling Approach

The decision about what transportation modeling approach to select for a work zone analysis is secondary to the decision to deploy a transportation modeling approach at all. However, once the decision is made to use a transportation analysis tool, the question turns to which tool is best suited given my circumstances? The difficult part is balancing the available resources (time, money, data, etc.) with the desired level of detail (e.g., ability to model one-lane/two way operations using historic traffic counts for a Type I project).

Given the fact that each work zone analysis will be different, developing a comprehensive checklist or decision-tree will never be adequate enough in selecting "the right tool". However, the case studies included in this document illustrate generally successful examples of managing technical risk as a part of an overall modeling approach. Using the Analytical Work Zone Decision Framework discussed previously, all of the 13 project application case studies and two of the strategic modeling case studies were placed on the framework to indicate how each ranked in terms of level of detail needed (low-medium-high) and available resources (low-medium-high). As shown in Figure 12, what the case studies reveal is a diverse mix of transportation modeling approaches based upon the level of detail needed and available resources.

From these case studies, a number of clusters can be created based upon the modeling approach chosen. These clusters are shown in Figure 13 and characterize the modeling approaches observed across the case studies based upon relative availability of resources and required detail for analysis. Figure 13 should not be considered specific recommendations but rather a reflection of the diversity of approaches developed by the analysts across the case studies. The transportation modeling approach clusters are highly consistent with the work zone modeling spectrum presented in Figure 1. As one moves along the modeling boundary bisecting the areas of significant technical risk and modeling opportunities, the complexity of the transportation modeling approaches increase. When both resource availability and the level of detail needed are low, the most appropriate transportation modeling approaches are likely to be sketch-planning and HCM methodologies. As resources increase, the available transportation modeling approaches increase and include macroscopic models for pipe network forms and travel demand models for those analyses with a grid network form. As the requirement for detail increases, the available transportation modeling approaches shift to mesoscopic and microscopic modeling tools. Mesoscopic applications include both pipe and network forms. Microscopic modeling tools are applied alone or as a part of a multi- scale approach, where the model is designed to be used strategically throughout the project life- cycle in conjunction with other tools.

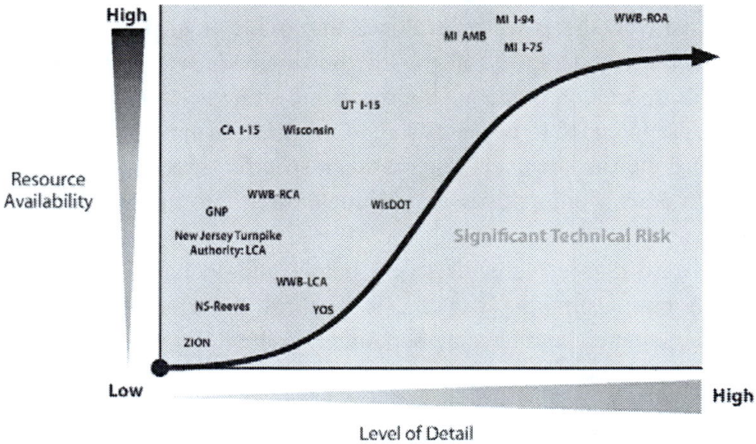

Figure 12. Analytical Work Zone Decision Framework—Case Studies.

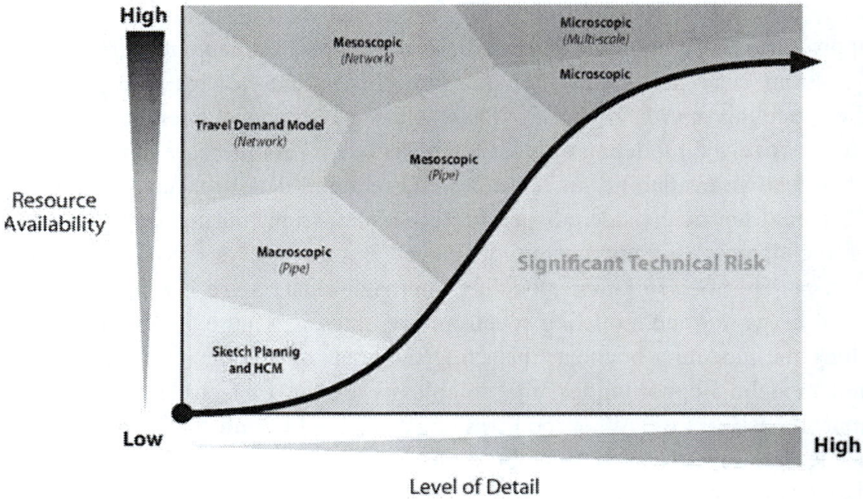

Figure 13. Analytical Work Zone Decision Framework—Modeling Approaches.

4.3. Developing a Transportation Modeling Approach

The preceding two sections established a strategic framework which can be used identify a specific transportation modeling approach. Some analysts, however, would like a tool to assist with identifying an appropriate transportation modeling approach and strategic methodology. To that end, the following two worksheets were developed to provide an analyst with a tool to conduct a detailed assessment of a work zone analysis project and ultimately identify a transportation modeling approach and strategic methodology. These worksheets are intended to be used as a *tool to guide* an analyst and should not be considered the final decision in choosing a specific transportation modeling approach. In other words, this is an additional input into the overall decision-making process.

The two worksheets are discussed below and include instructions for filling each out. Using the worksheets requires the analyst to have read through this document and have a good understanding of a work zone analysis project.

Work Zone Analysis Scoping Worksheet Instructions

The scoping worksheet should be used to document the various components of a work zone project that need to be analyzed using a transportation analysis tool. The scoping worksheet is based primarily upon Section 2.0, Work Zone Analysis Factors. The following instructions provide more detail on how to fill-in the scoping worksheet.

1. ***Key Decisions***—List the Key Decisions that need to be supported using the transportation modeling approach. References on the importance of articulating key decisions to be supported:
   - Section 1.2, Work Zone Analysis Decision Making
2. ***Work Zone Charcteristics***—Identify the Type, Network Configuration, Size, and Analysis Area Dimension of your work zone analysis project. Refer to Section 2.1 for more information on definitions.
3. ***Agency Resources***—Under each heading mark the following and refer to Section 2.4 for more information on definitions:
   a. *Institutional*: Circle those institutional categories which have been addressed by your agency.
   b. *Technical Staff*: Circle whether the technical expertise to run the transportation analysis tools resides in-house, via consultant or both.
   c. *Funding*: Indicate the amount of funding available.
      $ = $10,000 and $ = $100,000
   d. *Time to Analyze*: Circle the amount of time available to conduct the analysis
      short = Less than 2 months
      medium = 3 months to 6 months
      long = Greater than 6 months
4. ***Performance Measures***
   a. *Measures of Effectiveness*: Fill in each circle based upon the classification system indicated. Refer to Section 2.5 for more information and definitions.
   b. *Precision of Results*: Indicate on the bar whether the Key Decisions that need to be supported are either a rough estimation or a detailed assessment.
5. ***Data***—Indicate the following and refer to Section 2.3 for more information and definitions.
   a. *Type*: Indicate availability.
   b. *Source*: Indicate which data is available through archived or primary sources.
   c. *Quality*: Indicate availability.

6. ***Transportation Management Plan Strategies***—Indicate which transportation management plan strategies need to be included in the analysis based upon the classification system indicated. Refer to Section 2.2 for more information and definitions.
7. ***Fill-in Scoring Worksheet***

Work Zone Analysis Scoring Worksheet Instructions

The scoring worksheet should be used after the scoping worksheet has been filled in. The following instructions provide more detail on how to fill-in the scoring worksheet.

1. ***Develop a Transportation Modeling Approach***—This table should be filled in using the classification system indicated in the bottom left corner of the table. Each heading refers to a certain work zone factor from this document (Section 2.0) the analyst will need to reference in order to indicate whether the transportation modeling approach is appropriate. Based upon the scoping worksheet, the analyst should fill out each cell after reviewing the discussion and summary table for each section. The following provides a reference to the sections in the document:
   a. *Work Zone Characteristics*: Section 2.1 Work Zone Characteristics
   b. *Agency Resources*: Section 2.4 Agency Resources
   c. *Performance Measures*: Section 2.5 Work Zone Permance Measures
   d. *Data*: Section 2.3 Data
   e. *Transportation Management Plan*: Section 2.2 Transportation Management Plan Strategies
2. ***Summary***—Once each of the cells for each work zone factor is filled in, the analyst should provide an overall ranking (using the same classification system as in Step 1 above) for each transportation modeling approach. The analyst should rely on visual inspection of the results to indicate whether the transportation modeling approach is appropriate or not.

Work Zone Analysis Scoping Worksheet

What key decision are you supporting?

Work Zone Characteristics (Select One for Each Category)		Agency Resources	
Type	I / II / III / IV	*Institutional*	culture / leadership / data / contracting
Network Configuration	isolated / pipe / grid	*Technical Staff*	consultant / in-house
Size	small / medium / large	*Funding*	$$$$$$$$$$$$$
Analysis Area Dimension	site / local / metropolitan	*Time to Analyze*	short / medium / long

Performance Measures

● = Key Performance Measure; ⊘ = Secondary; ○ = Not Important

In the space adjacent to the circle, note any key policy requirements associated with the performance measure

Travel Time	○	*Volume*	○
Delay	○	*User Cost*	○
Queue Length	○	*Incidents*	○
Speed	○	*Cumulative Impacts*	○

Precision of Results rough estimation ●———● detailed assessment

Data

		Availability	Notes
Type	Roadway Characteristics		
	Travel Demand Characteristics		
	Cost Attributes		
Source	Primary		
	Archived		
Quality	Collection Frequency		
	Geographic Coverage		
	Archive Length		
	Accuracy		

Transportation Management Plan Strategies

■ Key TMP Strategy; ☑ Minor TMP Strategy; ☐ TMP Strategy Not Considered

Temporary Traffic Control			Public Info.	Transportation Operations			
Control Strategies	**Traffic Control Devices**	**Project Innovation**	**Motorist Information**	**Demand Mgmt Strategies**	**Corridor/Network Mgmt Strategies**	**Work Zone Safety Mgmt**	**Traffic/Incident Management**
☐ Construction phasing/ staging	☐ Changeable message signs	☐ Project Coordination	☐ Changeable message signs	☐ Transit service improvements	☐ Signal timing/ coordination	☐ Speed limit reduction/ variable speed limits	☐ ITS for traffic Monitoring
☐ Full roadway closure	☐ Flaggers and uniformed traffic control officers	☐ Contracting Strategies	☐ Dynamic speed message sign	☐ High occupancy vehicle lanes	☐ Temp. traffic signals	☐ Temporary traffic signals	☐ Transportation management center
☐ Lane shifts or closures	☐ Temporary traffic signals	☐ Innovative construction techniques	☐ Highway advisory radio	☐ Toll/congestion pricing	☐ Street/intersection improvements		☐ Tow/freeway service patrol
☐ One-lane, two way operations			☐ 511 traveler information systems	☐ Ramp metering	☐ Bus turnouts		☐ Local detour routes
☐ Two-way traffic on one side of divided facility					☐ Turn restrictions		
☐ Reversible lanes					☐ Separate truck lanes		
☐ Ramp closures/ relocation					☐ Reversible lanes		
☐ Freeway-to freeway interchange closures					☐ Ramp metering		
☐ Night work					☐ Temporary suspension of ramp metering		
☐ Weekend work					☐ Ramp closures		
☐ Work hour restrictions for peak travel					☐ Coordination with adjacent construction sites		
☐ Off-site detours/use of alternate routes							

Work Zone Analysis Scoring Worksheet

Develop a Transportation Modeling Approach

	Sketch-Planning and HCM Methodologies	*Travel Demand Models*	*Traffic Signal Optimization*	*Macroscopic Simulation*	*Mesoscopic Simulation*	*Microscopic Simulation*
Work Zone Characteristics						
Type						
Network Configuration						
Size						
Analysis Area						
Agency Resources						
Institutional						
Technical Staff						
Funding						
Schedule						
Performance Measures						
Travel Time						
Delay						
Queue Length						
Speed						
Volume						
User Cost						
Incidents						
Cumulative Impacts						
Data						
Type						
Source						
Quality						
Transportation Management Plan						
Control Strategies						
Traffic Control Devices						
Project Innovation						
Motorist Information						
Demand Mgmt Strategies						
Corridor/Network Management						
Work Safety Management						
Traffic/Incident Management						
SUMMARY						

● = Transportation modeling approach address work zone factor.
⊘ = Transportation modeling approach may or may not address work zone factor.
○ = Transportation modeling approach does not address work zone factor.

Identify a Strategic Methodology

Where are you in the project life-cycle?	
Planning	Multi-scale or Screening
PE/Design	Multi-scale or Screening
Construction	Mono-scale
Are there other key decisions that will need to be made?	
Yes	Multi-scale or Screening
No	Mono-scale
Is there uncertainty in the scale of the work zone type?	
Yes	Screening
No	Multi-scale or Mono-scale

Answer

Describe your transportation modeling approach below. Provide justification from the analysis above.

3. ***Identify a Strategic Methodology***—This section asks three basic questions regarding the work zone analysis to assist the analyst in

identifying one of the three Strategic Methodologies discussed in Section 3.0 Establishing a Strategic Methodology for Work Zone Analysis

4. ***Answer***—Once the three previous steps have been completed both a Transportation Modeling Approach and Strategic Methodology should become evident. The space provided in the Answer section should be used to describe and justify the analyst's reason for choosing each.

5.0. SUMMARY AND SYNTHESIS

The intent of this work zone modeling and simulation guide for analysts is to provide practical guidance on the identification and development of a transportation modeling approach for work zone analysis. The guidance is grounded in a detailed understanding of the key work zone factors in developing a transportation modeling approach (Section 2.0). This understanding is considered jointly with a strategic methodology for work zone analysis (Section 3.0). These two elements provide the foundation for a worksheet-driven exercise to assist the analyst in the development of transportation modeling approach for a specific work zone analysis project (Section 4.0). Throughout this process, a series of case study examples were used to illustrate how analysts have successfully identified modeling opportunities, mitigated technical risk, and conducted work zone impacts analyses to support key decisions across a range of considerations, including constructability, scheduling and TMP design, and evaluation.

What the preceding discussion and case study examples demonstrate is that work zone decision- making is an evolutionary process. Any work zone analysis should be linked to a specific decision with the knowledge that decisions made early in the project life-cycle have a large impact on future decisions. Placed within the context of analyzing work zone impacts, the further along in the project life-cycle one is, the lower the relative opportunity for transportation models to have a role in supporting the decision-making process. As seen in Figure 14 below, the decision-making evolution process combines the work zone decision-making process along with the work zone decision-making engine. Below it, represented by two polygons, are the tradeoffs that are made between modeling opportunities and data requirements as the decision-making process evolves over time.

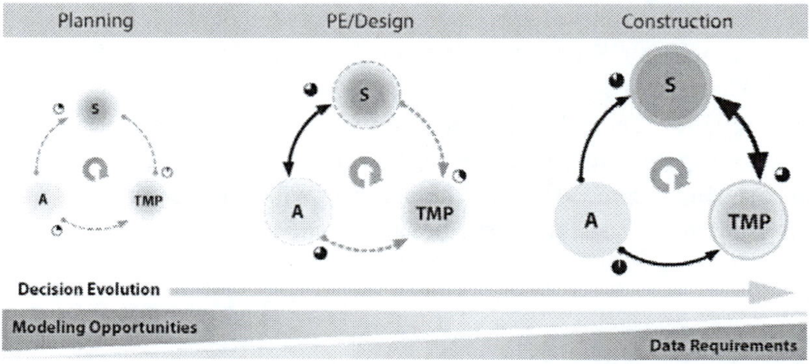

Figure 14. Work Zone Decision-Making Engine Process.

In the end, the options available to an analyst in deploying a transportation modeling approach decrease as more decisions are made regarding application techniques, scheduling, and TMP strategies. For example, once planning is complete and the decision to conduct a full-closure has been made, the use of a transportation modeling approach to help mitigate the mobility impacts is limited in that the decisions to be supported are much narrower. However, as seen in many of the case studies, the use of a transportation analysis tool may still be warranted regardless of the project life-cycle stage.

5.1. Key Points

TAT Volume IX highlighted four key points that are important to consider when an analyst is deciding upon which transportation modeling approach to ultimately develop and use:

- ***Work zone decision-making is an evolutionary process which supports a number of key questions to be answered.***—As seen in Figure 14 and discussed throughout this document, work zone analysis evolves over time as the roadway construction project moves forward from Planning to PE/Design to Construction and key decisions need to be supported. For example, during the planning stage, the work zone analysis may focus on numerous TMP alternatives and the key question is centered upon which alternatives are feasible given certain criteria. Later, a new set of questions may need to be answered that places more constraints on the feasible list of

TMP alternatives (e.g., available funding or political pressure). Thus, as the overall project moves forward it is always evolving and the use of a transportation modeling approach is always predicated on certain key decisions that have to be made.

- ***Data requirements are a driving force behind developing a transportation modeling approach***—Section 2.0 provides a detailed discussion of the many different factors that need to be considered when selecting a transportation modeling approach. For example, selecting an approach may be constrained by agency resources in terms of budget and staff resource. Or, the tool selection may be driven by the utilization of a cutting-edge TMP strategy such as variable speed limits through a work zone which requires a detailed vehicle-based evaluation. Regardless, it is a complex issue to unravel. However, what each of these examples shows is that data requirements are an important factor to consider. As seen in Figure 14 above, data requirements increase as the project evolves from planning to construction thereby placing constraints on the use of a transportation modeling approach. Thus, as the project evolves and different questions need to be answered, so too will different data need to be used/collected which may create both constraints and opportunities in developing a transportation modeling approach.
- ***Identifying a Strategic Methodology early in the project life-cycle can mitigate future work zone analysis problems***—As discussed in Section 3.0, three strategic methodologies were presented representing various modeling opportunities for work zone analysis: mono-scale, screening, and multi-scale. These methodologies serve as a foundation on which transportation modeling approaches could be incorporated into the overall work zone decision-making framework. As shown in the Figure 14, the opportunity associated with choosing a transportation modeling approach is much greater early on in the project life-cycle. For example, during the planning stage relatively few decisions have been made and the availability and use of a transportation analysis tool are fairly large. This is in contrast to the construction stage at which time the application technique, scheduling, and TMP have been finalized and the opportunity to select a transportation analysis tool and for it to generate useful information are much more limited. However, by identifying a strategic methodology in the beginning of the project life-cycle, the work zone analysis opportunities do not necessarily decrease over time but rather

grow as the model can be leveraged and applied to more work zone analyses as warranted or needed.

- ***Developing a transportation modeling approach is unique to each work zone analysis***— While many people desire a prescriptive "cookbook" method to direct them in determining which transportation modeling approach to use, making that determination based solely upon one particular factor is problematic. What this document demonstrates is that there are many factors that go into developing a transportation modeling approach. For example, the Woodrow Wilson Bridge, a Type I work zone, used both simple HCM methodologies and complex microscopic simulation tools. Thus, the use of the worksheets developed in Section 4.0 are intended as a tool to guide analysts and provide a framework with which to better conceptualize and address the complexities associated with conducting a work zone analysis. Some agencies have developed specific procedures to guide the analyst in the development of an appropriate modeling approaches (such as the Wisconsin DOT Transportation Management Plan Development Process discussed in the case study), and these systematic approaches can both speed the modeling approach development process as well as ensure consistency when different staff consider projects independently.

5.2. Where Do I Go from Here?

Ultimately, the questions for many analysts boils downs to which transportation analysis tool should I use? Where do I get the data from? And, how do I overcome my institutional issues? These are difficult questions to answer and it is not the intent of this document to specifically address them given that each circumstance is unique. However, the following provides some direction on where one can turn to for help.

- ***Individual Transportation Analysis Tool Resources***—It is not the intent of this document to provide guidance on the use or application of any one specific transportation analysis tool. Rather, this guidance is intended to help an analyst identify a certain transportation modeling approach for work zone analysis. There are many unique transportation analysis tools available under each modeling approach. Often, these tools are constantly changing as capabilities are added as well as being routinely updated and tweaked. Also, new tools become available on a regular basis. Recently, many commercial vendors have begun offering hybrid models that span more than one class of tools such as mesoscopic models that also operate on the microscopic level. For more information on using an individual transportation analysis tool refer to the following:

- **FHWA Traffic Analysis Toolbox (TAT)**: The Traffic Analysis Tools Program was formulated by FHWA in an attempt to strike a balance between efforts to develop new, improved tools in support of traffic operations analysis and efforts to facilitate the deployment and use of existing tools. The TAT provides a detailed description of traffic modeling tools. http://ops.fhwa.dot.gov/trafficanalysistools/index.htm.
- **McTrans**: Centeter for Microcomputers in Transportation (McTrans) serves as a resource for the distribution and support of microcomputer software in the highway transportation field. McTrans provides expert technical advice, information exchange and a wide range of transportation-related software at very reasonable costs. http://mctrans.ce.ufl.edu/

- ***Data Collection***—Data collection is routinely cited as a barrier to using transportation modeling. As the expression goes: "garbage in, garbage out". With the increasing use of sensors located throughout the transportation system, the ability to cost effectively collect traffic data has increased. Often, much of the required data has probably already been collected and simply requires contacting another department within the agency. Some examples are as follows:
  - Many transportation agencies have permanent count stations on Interstate and arterial roadways from which traffic count data is routinely collected, cleaned, and archived. A case in point is the Minnesota Department of Transportation Traffic Forecasting & Analysis unit which makes its data available on its website (http://www.dot.state. mn.us/traffic/data/html/traffic.html).
  - Another opportunity for data collection is the use of portable traffic counters. The Federal Lands Highway Division purchased a number of counters for the sole purpose of data collection for transportation modeling. See the Yosemite National Park: Yosemite Village Roadway Reconstruction case study for more information.

- For major roadway reconstruction projects the use of real-time monitoring may yield useful data. This occurred as part of the Woodrow Wilson Bridge project where a traffic data collection effort was undertaken to establish a formalized process for collecting and archiving traffic data as part of the project for use in future studies and analyses. See the Woodrow Wilson Bridge Reconstruction case study for more information.

- ***Addressing Institutional and Policy Issues***—Institutional polices can often be a significant driver in developing a transportation modeling approach. The discussion on Institutional Arrangements (Section 2.4.1) lists four areas general areas that can impact transportation modeling: culture, leadership, data management, and contracting. Many progressive transportation agencies have addressed these areas and the FHWA has available numerous resources addressing the subject. Another institutional area that needs to be addressed is work zone policies that have been created. For example, many agencies now have restrictions on lane closures in work zones, available work hours, maximum user delay, etc. These policies, created by the transportation institutions, have an impact on the overall usefulness of a transportation modeling approach. For example, a policy limiting the maximum user delay may negate the results of a model that demonstrate a larger maximum user delay for a shorter duration is a better choice overall for a construction project. While many of these policies are developed with good intentions, the use of a transportation analysis tool may contribute to the development a more refined approach. For more information on work zone policy issues refer to the following FHWA material concerning the Final Rule on Work Zone Safety and Mobility:
  - **Implementing the Rule on Work Zone Safety and Mobility** (HTML, PDF)— Provides a general overview of the Rule, as well as guidance, examples, best practices, tools, and resources to help implement the Rule's provisions.
  - **Work Zone Impacts Assessment: An Approach to Assess and Manage Work Zone Safety and Mobility Impacts of Road Projects** (HTML, PDF)—Provides guidance on developing procedures to assess and manage work zone impacts of road projects, as well as examples and practices of how agencies are currently assessing and managing work zone impacts.

- **Developing and Implementing Transportation Management Plans for Work Zones** (HTML, PDF)—Provides information about developing and implementing Transportation Management Plans (TMP), including information on how and where a TMP fits into project-level processes and procedures, a list of components that can be considered for inclusion in a TMP, descriptions of work zone management strategies, and examples and practices of how agencies are currently using TMPs.
- **Work Zone Public Information and Outreach Strategies** (HTML, PDF)—Provides tips, examples, and practices on designing a public information and outreach campaign for work zones and offers a variety of strategies that can be used in a campaign.

CASE STUDIES

Caltrans I-15 Pavement Reconstruction Project

Overview

The I-15 Pavement Reconstruction project (began construction in October 2008) will improve a four-mile stretch of I-15 near the interchange with I-10 in Ontario, California (referred to as the I-15 Ontario corridor). The reconstruction will rebuild two to three truck lanes in each direction, replacing damaged concrete slabs and base pavements with new concrete slabs using Rapid Strength Concrete (RSC) and new Asphalt Concrete (AC) base. Currently, this section of I-15 includes four to six lanes in each direction and carries an AADT of approximately 200,000, with a particularly high percentage of heavy truck traffic (about 12 percent on average on weekdays). In addition, the I-15 Ontario corridor has consistently high weekday commuter peaks and similar volumes on weekends due to leisure travelers from Los Angeles headed to and from Las Vegas and resort locations along the Colorado River. The project scope and study area are shown in Figure 15.

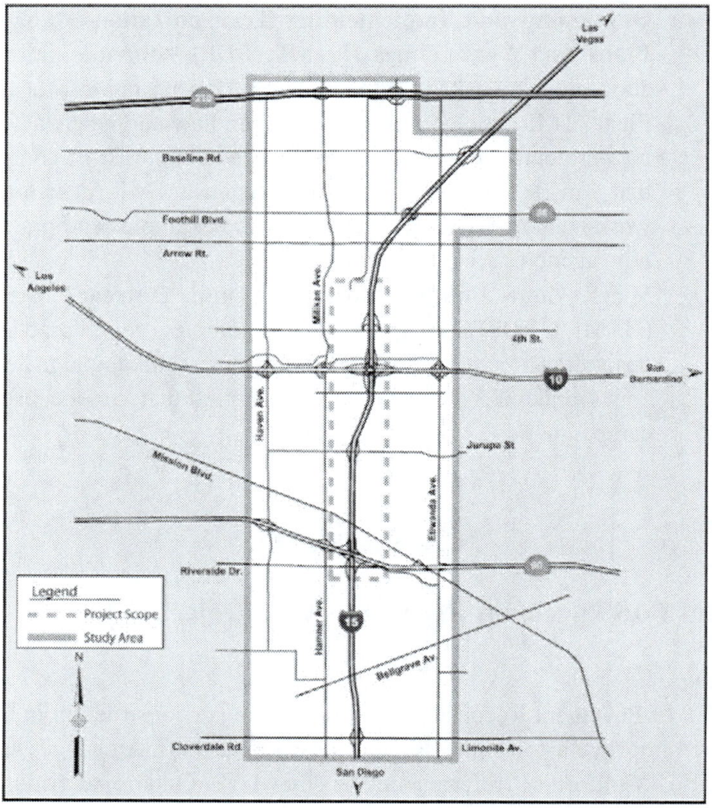

Figure 15. I-15 Pavement Reconstruction Analysis Area.

California Department of Transportation I-15 Pavement Reconstruction Project[4]

Work Zone Characteristics	
Transportation Analysis:	
Approach	Sketch-Planning Mesoscale
Modeling Tools	CA4PRS HCM DynamEq
Work Zones:	
Type	Type II
Network Configuration	Pipe
Geographic Scale:	
Work Zone Size	Medium

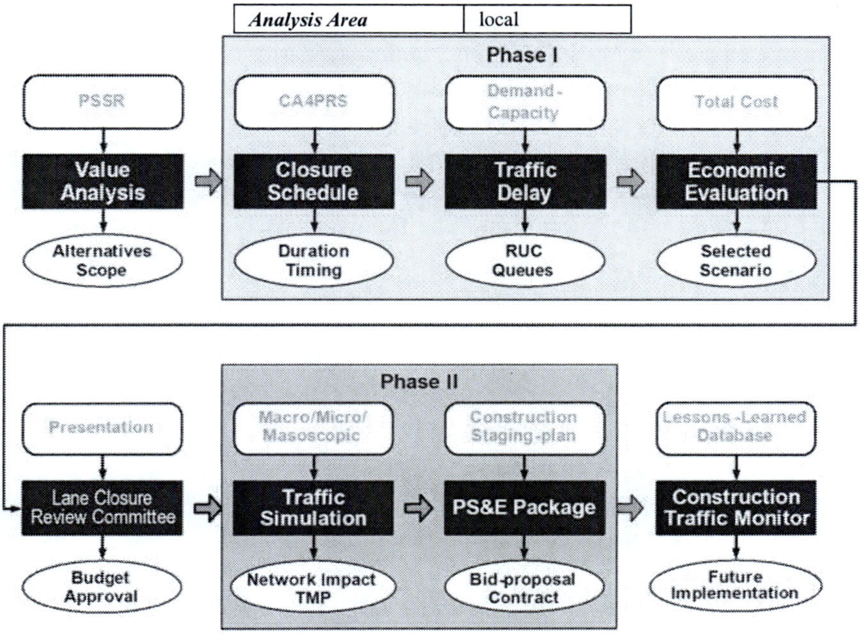

Figure 16. I-15 Analysis Process.

Caltrans gave the preliminary notice to proceed with the design and analysis of the project in December 2004 after an extensive planning process. During the design stage, the integrated pre- construction analysis was to help Caltrans compare all feasible scenarios for the I-15 Ontario project and select the best approach in terms of *production schedule*, *traffic delay*, and *total cost*. To that end, Caltrans employed the use of three traffic modeling tools including the Construction Analysis for Pavement Rehabilitation Strategies (CA4PRS), Highway Capacity Manual Demand-Capacity Model (HCM), and the DynamEq mesoscopic traffic network analysis tool. The use of CA4PRS and HCM represent Phase I of the preconstruction analysis whereby six alternative construction scenarios were analyzed. The use of DynamEq represents phase II of the pre-construction analysis which entailed the development of traffic management plans. This process is shown in Figure 16.

The following sections will provide more detail on the use CA4PRS, HCM and DynamEq in the two phases of the pre-construction analysis.

Phase I: Preconstruction Analysis

In the Phase I analysis, six construction scenarios were developed and analyzed using the CA4RPS tool in order to compare number of closures and the HCM to quantify queues and delay costs. The results of these two traffic models were included in an overall cost estimate analysis to determine the most appropriate construction scenario in terms of construction scheduling, impact on drivers and project costs. The following six construction scenarios were analyzed:

- Median widening (Original scheme)
- Value Analysis (VA) By-pass (eventually eliminated from overall analysis)
- Rapid Rehab (55-hour Weekend)
- Rapid Rehab (Continuous Progressive)
- Traditional nighttime (PCC slab replacement)
- Crack-Seat and AC overlay

Closure Assessment: CA4RPS—The CA4PRS was employed because it easily evaluates "what if" scenarios for highway rehabilitation in order to identify solutions that balance on-schedule construction production, traffic inconvenience, and agency costs. In addition, the CA4PRS results are easily integrated with various traffic simulation tools for quantifying road user costs during construction. This feature helped the project team determine rehabilitation strategies that maximize production and minimize inconvenience to the public. Caltrans collected data for the I-15 CA4RPS analysis using previous analyses of similar projects from around the state including I-10 Pomona, I-710 Long Beach and I-15 Devore. Results of the CA4RPS included the total number of closures as listed in the column labeled *Number of Closures* in Table X.

Traffic Analysis: HCM— The impacts of construction work zone closures analyzed using the CA4RPS on freeway traffic were quantified using the HCM Demand-Capacity Model. The model was created in Microsoft Excel based upon the methodology articulated in the HCM. First, total and maximum delays were calculated with variable traffic demand reduction by percentage and work zone roadway capacity. Second, road user cost (the time value of travelers adjusted by an annual growth factor) was estimated based on the calculated vehicle-delay hours during construction for each scenario. Finally, the California Freeway Performance Measurement System (PeMS) was used to estimate traffic demand. The results of this analysis (which included various sensitivity analyses) included two values: road user cost (RUC) and maximum roadway user delay per person. These results are listed in Table 6.

Cost Analysis—A typical construction cost analysis was conducted for each of the scenarios that did not include any type of traffic modeling component.

Results from the Closure Assessment, Traffic Analysis and Cost Analysis are listed in Table 6. As can be seen in the results, the outputs from the Closure Assessment, Traffic Analysis and Cost Analysis were used to develop an overall Total Cost. The Closure Assessment values were used to estimate the RUC and Delay for the Traffic Analysis. The RUC was discounted by one-third and added to the Cost Analysis value to create a Total Cost. These results indicate that Scenario 1, Original, is the most cost-effective after considering all of the critical variables involved.

Phase II: DynamEq

The Phase II analysis for the I-15 Ontario project entailed two sub-studies: (1) a traffic network simulation using DynamEq to assist in the development of a TMP; and (2) a supplemental construction staging plan analysis which involved a more detailed analysis using the CA4RPS tool. Of primary concern here is the use of the DynamEq traffic modeling tool for developing the TMP. Based upon the Phase I Preconstruction Analysis, the recommended construction scenario was number 1 which entailed a total of 35 55-hour long weekend road closures. DynamEq was specifically used to analyze the detailed Construction Staging Plan and conduct an operational analysis of the primary detour routes for six key stages in the construction process where detours were considered to be critical. The six staging scenarios analyzed are summarized below:

- 2B—Closure of I-10W to I-15S and I-15S to Jurupa ramps
- 2C—Closure of I-10E to I-15S ramp
- 2D—Closures of I-15S to I-10W and from 4th Street to I-15S ramps
- 3D—Closures of Jurupa to I-15N and I-15N to I-10E and I-10W ramps
- 3F—Closure of EB I-10 to NB I-15 connector and both NB ramps at 4th Street
- 4B—Closure of SB I-15 to WB and EB SR 60

Table 6. I-15 Reconstruction Assessment Result

Scenario Name	Closure Scheme	Closure Assessment (Number of Closures)	Traffic Analysis		Cost Analysis	Total Cost[5]
			RUC ($M)	Delay (Min)		
Original	Median + structure widening	35 weekends	3	16	78	**79**
Rapid Rehab 1	Full closure one roadbed	35 weekends	119	363	83	**123**
Rapid Rehab 2	Full closure one roadbed	8 weeks	123	363	77	**118**
Traditional	Partial closure	1,220 nights	133	22	88	**113**
Long-life CSOL	Full closure one roadbed	20 weekends	69	363	60	**83**

DynamEq was employed for use in this analysis due to the relatively large study area and the number of detour routes involved. While DynamEq is a user equilibrium model (meaning vehicles make optimal use of the network with the goal of minimizing overall system delay) the reality is that many of the vehicles operating on the network are unfamiliar to the area and do not have perfect information. Thus, there will be in-equilibrium in system usage. In order to account for this behavior in the DynamEq model, a new class of vehicles was created that are unfamiliar with the area and will always follow posted detour routes regardless of congestion levels.

Results of the study indicated significant user delay ranging from 8 to 121.4 minutes. Four of the six staging scenarios had significant delays of more than 30 minutes. The analysis results had two major impacts on the overall project design and development of the TMP. First, the results indicated the need for more than one detour route for some of the staging scenarios. This problem was not previously identified in other analyses. Second, a public outreach campaign is critical to ensuring demand reductions and detour routing.

Glacier National Park: Going to the Sun Road Rehabilitation Project

Overview

The Going-to-the-Sun Road (GTSR) in Glacier National Park (GNP), Montana, is a prime attraction for visitors to the park and the only east-west link within the park. The scenic 50-mile roadway, completed in the 1930s, traverses the park and provides access to the Logan Pass Visitors Center from either the St. Mary's Entrance (East) or the West Entrance. While portions of the GTSR are open throughout the year, in the higher alpine sections, the roadway is closed and often snow- covered throughout the winter months. The roadway offers the visitor a number of incredible vistas as well as access to trailheads and other facilities along its length. Given the rugged terrain, the GTSR is the only roadway within the park that connects the eastern and western sections of the park. Alternative east-west connections outside the park do exist, however, the GTSR itself or one of the facilities along its length are the destinations of the vast majority of vehicles visiting the park.

The GTSR is slated for an extensive multi-year rehabilitation project over a seven to eight year period. Since the GTSR is a key reason for visiting GNP and has no alternative route, the roadway must remain open throughout the project to both allow access to facilities and to remain open as a scenic roadway with minimal disruption and delay for park visitors. The steep terrain, the complexity and duration of the work to be performed, and the limited construction season for roadwork in the summer season (coinciding with peak visitor travel demand) are factors that need to be considered in the planning stage to determine if the GTSR Rehabilitation Project can be completed in a timely manner while still maintaining an acceptable level of delay to visitors using the GTSR.

Because of the key role the GTSR plays within the park as a destination itself and in providing east- west access in the park, there was significant concern that extended and onerous delays from work zones would impact park visitation as well as the segments of the local economy that depended on tourism. An agreement was reached between the National Park Service, Western Federal Lands Highway Division (WFLHD) and representatives of the local community outlining the extent to which closures and delays on the GTSR would be tolerated. One key tenet of this agreement was to limit end-to-end delays on the GTSR in one direction to no more than 30 minutes (total).

Glacier National Park
Going to the Sun Road Rehabilitation Project[6]

Work Zone Characteristics	
Transportation Analysis:	
Approach	Sketch-Planning
Modeling Tools	QuickZone
Work Zones:	
Type	Type III
Network Configuration	Pipe
Geographic Scale:	
Work Zone Size	Large
Analysis Area	local

The role of FLH-QuickZone in the GTSR Rehabilitation Project was to assess likely travel delays (and in particular, the end-to-end delays from multiple work zones) expected over the course of the multi-year project. Since the project was, at the time, in the relatively early planning phase, the details of roadwork phasing and staging could not be identified in detail. However, given the outline of a likely phasing plan and expected traffic control developed by WFLHD, FLH- QuickZone was used to identify projected delays and queue length over the course of the project. FLH-QuickZone was also employed to assess the likely effects of actuated signal control for 2-way, 1-lane operations, as well as the impact of reduced travel demand. Reduced travel demand is projected during roadwork, as well as a result of the provision of a transit alternative.

Work Zone Alternatives

Eight alternatives were coded and analyzed using FLH-QuickZone for the GTSR Rehabilitation Project (see Figure 17). The eight alternatives (Table 7) are combinations of four expectations for travel demand and the utilization of actuated signal control for nighttime and weekend 2-way, 1-lane operations. The reduction in demand is from two potential sources. The first reflects an estimate in the EIS that the presence of major roadwork will cause overall GTSR travel demand to decline by approximately 6 percent.[7] The second is a planning level target by planners to reduce travel demand by 2 to 7 percent by shifting visitors into transit buses. The 6 percent demand reduction case assumes just the overall GTSR decline from the presence of roadwork. The -8 percent and -13 percent demand reduction cases reflect a combination of general decline with the high and low estimates of transit demand shift.

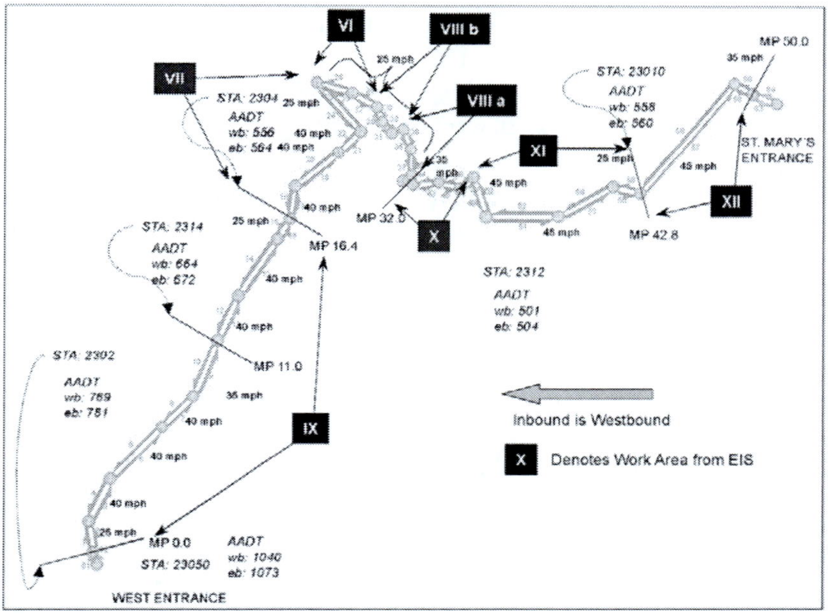

Figure 17. GTSR Rehabilitation Project.

Table 7. GTSR Alternatives Evaluated with FLH-QuickZone

			Work Zone Traffic Control	
#	Alternative Name	Travel Demand	Weekday Day 7am-7pm	Weekend (all hours) and Weekday Night 7pm-7am
1	Base	2004 Level	Flaggers	Fixed Time Signals
2	Actuated Signal	2004 Level	Flaggers	Actuated Signals
3	-6% Demand	-6%	Flaggers	Fixed Time Signals
4	-6% Dem + Actuated	-6%	Flaggers	Actuated Signals
5	-8% Demand	-8%	Flaggers	Fixed Time Signals
6	-8% Dem + Actuated	-8%	Flaggers	Actuated Signals
7	-13% Demand	-13%	Flaggers	Fixed Time Signals
8	-13% Dem + Actuated	-13%	Flaggers	Actuated Signals

The use of a single fixed timing plan for the control of short work zones has been a typical practice on GTSR, as is the use of varying fixed timing plans over the course of the day for longer work zones. In the Glacier case study, WFLHD and NPS staff wanted to investigate the complete range of delay from the simplest (single fixed plan) to the most complex (all actuated) and to estimate how likely the 30 minute maximum user delay threshold might be exceeded over the life of the rehabilitation project. Base signal timing in this case study represents a single fixed plan used in the night and weekend periods when flaggers are not present.

Application

For the FLH-QuickZone analysis of the GTSR project, there were two critical measures of effectiveness to consider, both related to the designation of 30 minutes or more of delay in either direction as 'unacceptable" throughout the project. In order to describe the worst delay seen in a phase, *maximum user delay* was used.. This measure reflects the longest possible delay on the GTSR from all work zones encountered in one direction. During 2-way, 1-lane operations, the assumption implies that some unlucky traveler will arrive at a work zone to experience the longest wait possible in that time period. The second measure of effectiveness is the number of *hours per week that the maximum delay exceeds the 30 minute threshold* in one or more directions. This measure provides insight into how long unacceptable delays are in effect throughout the week.

Base Case--In general, the FLH-QuickZone results indicate that delays in the base roadwork case are frequently in the unacceptable range, and remain in that range for a significant portion of the week for several phases. In the base case, the delays are particularly noteworthy during night operations and in specific high-volume (July) time frames. During high-volume time frames, flaggers are not expected to be able to allow all cars in the queue to pass through before a maximum time of 30 minutes is reached. The result is oversaturation and queue development. This development is relatively short-lived, however—no delay exceeds 44.9 minutes.

During night operations, delay is high because a single fixed plan must be used to cover high demand weekend patterns as well as low-demand night patterns. The result is that relatively long cycle lengths are imposed to prevent saturation on weekends, but cause long waits at red lights at night (up to roughly 12 minutes). It is also clear that 12 minute wait times under low flow conditions may be unsafe since the road user may presume that the signals are broken after a few minutes, proceed and eventually encounter oncoming traffic in the single open lane of the work zone. Before such unrealistic situations were to actually occur, it is likely that timing plans that change periodically throughout the evenings and weekends would be put into place. This option was not analyzed in QuickZone-FLH as a part of this study at the direction of WFLHD and NPS staff because of a lack of time to explore every possible variant within the limitations of the case study. Given the preliminary nature of the case study, WFLHD and NPS staff wanted to concentrate effort on identifying the likely worst case and best case delay outcomes.

Signal Actuation—The impact of signal actuation is significant in the reduction of delays during night operations. By allowing the signal timing to vary with actual travel demand, more appropriate (short) cycle lengths can be provided at night, and longer ones during weekend and weekday evenings (after flaggers are finished for the day). Results indicated that the use of actuated signals eliminates unacceptable delays at night, although delays up to 25.3 minutes are predicted. Signal actuation alone is not enough to eliminate unacceptable delays during the July and Late Summer peak, however, particularly in 2011 and later. Here, travel demand exceeds the capacity of the expected work zones, regardless of signal timing.

Reduced Travel Demand—The reduction in demand is effective in reducing daytime delay during flagger operations but has no effect on night operations when fixed timing plans are in place. Results indicated that this effect is highest with the -13 percent demand case and somewhat lower in the other two cases. Overall, the demand reduction is only a critical factor when the highest seasonal travel demand is expected: July. In other months, the delay from the flagger operations is often at acceptable levels. Actuated signals, in combination with reduced demand (-13 percent), eliminates unacceptable delays in all phases except for July 2011, when five concurrent work zones are in place.

Table 8 provides a summary of both the fixed signal timing plan impacts and the actuated signal impacts respectively. Again, using fixed signal timing plans results in unacceptable delay in more than half (21) of the nighttime FLH-QuickZone phases. Switching to actuated signals eliminates all delays of greater than 30 minutes. Switching to actuated signals during the daytime FLH- QuickZone phases helps to reduce the occurrences of unacceptable delays but does not eliminate them as was done with the nighttime operations. Nor does the use of actuated signals reduce the severity of the delay. The minimum and maximum duration of the delay during daytime operations remains around 35 minutes with actuated signals. Other mitigation measures beyond those studied here, or a reduction in work zone activity in the peak of the summer visitation season, will likely need to be considered if all potential instances of user delay exceeding 30 minutes are to be eliminated.

Based on this planning-phase analysis of the GTSR, it is clear that Alternative #1 will result in unacceptable delays on a regular basis during both nighttime and day work zone operations. The predicted value of actuated signals is large given the high variability of travel demand expected during the course of the project. If actuated signals cannot be reliably implemented, it is possible that a fixed plan that varies by time of day could also be implemented to mitigate delays, although such a plan has not been evaluated in FLH-QuickZone. The case for demand reduction is less clear – the problem lies primarily with reducing (or potentially redistributing) travel demand in the peak month of July, rather than in all time frames.

Table 8. Fixed and Actuated Signal Timing Impacts

Signal Operation		Number of QZ Phases		Hours/Weeks > 30 min Delay (hours)				Delay (minutes)			
				Low	High	Low	High	Min	Max	Min	Max
		Fixed	Act.	Fixed		Act.		Fixed		Act.	
Base Alternative	Day	8	5	3	40	7	35	30.2	44.9	31.1	36.3
	Night	21	0	48	48	0	0	30.3	37	0	0
-13% Demand Alternative	Day	1	1	–	30	–	30	–	32.4	–	32.4
	Night	21	0	48	48	0	0	30.3	37	0	0
-6% Demand Alternative	Day	4	3	8	30	20	30	32.3	36.1	30.5	34.3
	Night	21	0	48	48	0	0	30.3	37	0	0
-8% Demand Alternative	Day	4	3	5	30	10	30	30.6	33.7	30.2	33.7
	Night	21	0	48	48	0	0	30.3	37	0	0

Maryland State Highway Administration—Lane Closure Analysis Program

Maryland State Highway Administration Lane Closure Analysis Program[8]

Work Zone Characteristics	
Transportation Analysis:	
Approach	Sketch-Planning
Modeling Tools	Lane closure analysis program built with Microsoft Excel
Work Zones:	
Type	Type II – IV
Network Configuration	Isolated
Geographic Scale:	
Work Zone Size	Small
Analysis Area	Local

Overview

The Maryland State Highway Administration (SHA) has developed the Lane Closure Analysis Program (LCAP) to support state traffic engineers with a structured method to analyze work zone impacts. The LCAP tool was developed based upon the guidance written in the Work Zone Lane Closure Analysis Guidelines which present allowable thresholds for decreasing mobility measured in terms of queues and delays in work zones. LCAP is part of a tiered approach to work zone analysis that SHA has developed. Depending on the complexity of a project, SHA recommends starting with LCAP for simple cases, QuickZone for simple cases with network impacts, and then simulation tools such as CORSIM, VISSIM, and SimTraffic for relatively complex scenarios requiring detailed analysis.

LCAP Description

LCAP is an analytical tool designed to quantify queues and delays resulting from capacity decreases in freeway work zones. LCAP is written as a program within Microsoft Excel and compares expected travel demand against work zone capacity on an hour-by-hour basis to estimate delay and mainline queue growth.

LCAP consists of a four-step process:

Step 1: *Open existing file or Start new file.* The user has the choice of opening an existing LCAP file or starting a new case for analysis.

Enter project information. The Project Information screen allows the user to input data that defines the project. The project description, analyst's name and date of analysis should be input here. The project description should contain information to identify the project, such as roadway name, direction of travel, segment of road, and work being performed.

Step 2: *Input demand.* The demand is an essential part of the LCAP program. Users may enter their own data or access SHA's traffic data from the Traffic Monitoring website (http://www. marylandroads.com/tmsreports/).

Step 3: *Describe the work zone.* The closure analysis screen is used to determine the hours during the day when lanes can be closed without causing excessive queuing or delays. The user is prompted to select input options in terms of maximum allowable queue length or maximum allowable delay (to be highlighted in red or yellow when exceeded) and proposed lane closure hours. Using trial and error in the start and finish times of the work zone, lane closure schedules can be determined based on allowable queues or delays.

- *Lane Closure Information*—The user enters the total number of freeway lanes and the number of open lanes with the lane closure in effect.
- *Capacity Information*—The user enters the roadway capacity without and with the work zone in vehicles per hour per lane. The work zone capacity can be determined in four ways:
    1. engineering judgment,
    2. 1997 Highway Capacity Manual,
    3. University of Maryland Capacity Estimation equation, or
    4. 2000 Highway Capacity Manual equation.
- *OutputType*—The user selects queue length or delay as the measure of the work zones mobility impact. The user has the option of highlighting the output in red or yellow if the queue exceeds a specified length (miles) or the delay exceeds a specified period of time (minutes).

Step 4: *Save File.* LCAP has the ability to save all of the data associated with the project.

Michigan Department of Transportation—Southeastern Michigan Simulation Network (SEMSIN)

Overview

In 2005, the Michigan Department of Transportation (MDOT) was involved in the environmental clearance process for several major projects dealing with long-distance interstate and international trade. Since each of these projects would have major impacts on network freight flows, especially on the Detroit area freeway network, there was a need to analyze the individual projects within the framework of the larger system and in relation to each other. In addition, much of the Detroit area freeway system operates at or near capacity. Traditional travel demand models and highway capacity analysis could not fully capture the dynamic effects of these impacts on the network. Lastly, divisions of the MDOT responsible for operations (for example, work zone mobility) needed a traffic tool that would enable them to examine the network effects of major road closures and/or detours. Thus, the MDOT saw a need to develop a large-scale, network simulation model, capable of modeling passenger cars, domestic trucks and international trucks, and which would support planning and operational decisions. An SPR-funded pilot project was initiated to develop a network a simulation model. Initially, the core of the Detroit freeway network would be modeled, plus the I-75 corridor extending from the City of Detroit in Wayne County, to the industrial heart of Oakland County, a distance of 30 miles. This project was called the Southeastern Michigan Simulation (SEMSIM).

However, before the SEMSIM, which was a pilot study, was completed, there was an immediate need by three other very large projects for network simulation. The SEMSIM, in its then current state of development, served as the platform, or seed model for these three projects. Specifically, each project, The Ambassador Bridge Gateway Project, the I-94 Rehabilitation Project, and the I-75 Trade Corridor Project, cut out large sections of the SEMSIM network, extending freeway links and adding surface arterial corridors as needed.

The SEMSIM network was created using the Paramics microscopic simulation model and is show in Figure 18 on the left. The network incorporates the Central Detroit Freeway Network and the I-75 corridor extending to Pontiac, Michigan. The overall size is approximately 30 miles north/ south and 15 miles east/west. It includes all freeways and most major arterial roads. On the right are the locations of three maintenance of traffic simulation (MOTSIM) assessments that have been or are being conducted based upon the SEMSIM network. These three projects include the Ambassador Gateway Bridge MOTSIM (red), I-75 Trade Corridor MOTSIM (Purple), and the I-94 Rehab MOTSIM (green). The following sections will describe in more detail the application of the SEMSIN network for the three MOTSIM projects.

Ambassador Gateway Bridge MOTSIM

The first application of the SEMSIM network was the rehabilitation and reconstruction of the I-96/ I75 interchange adjacent to the Ambassador Bridge that provides the major commercial connection between Canada and the United States (Figure 19). The overall project includes reconstructing and realigning a 1.5 mile segment of I-75 and 1 mile segment of I-96, and construction of a new interchange with direct access ramps for trucks to the to the Ambassador Bridge, which is the largest commercial border crossing in North America. The reconstruction includes the full closure of many sections of the elevated freeway near the interchange which provides critical connections for freight movement both across the boarder and within the Detroit area. The project began in February 2008 with an estimated completion date of December 2009. A contractor incentive bonus of up to $8 million is being offered for early completion. In all, three separate analyses were conducted as part of the project during the construction stage:

- **Gateway MOTSIM 1**—Initial assessment which included surrounding roads near the construction project and interstate and arterial roads leading into and around the construction project (red boarder in Figure 2).
- **Gateway MOTSIM 2**—An expansion of the MOTSIM 1 network to include proposed lane closures for the Summer 2008 construction season along I-75 to the north and south (highlighted purple sections in Figure 2).
- **Gateway MOTSIM 3**—Developed in March 2008 based upon actual traffic conditions occurring as part of the full closure that began in February 2008. This analysis included a further assessment of Gateway MOTSIM 2 based upon actual conditions.

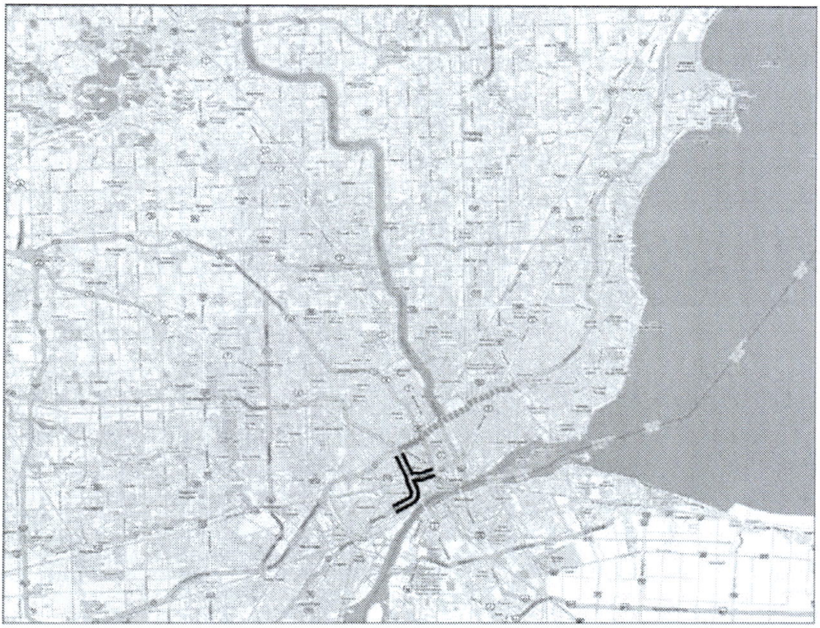

Figure 18. MDOT METSIM Geographic Coverage.

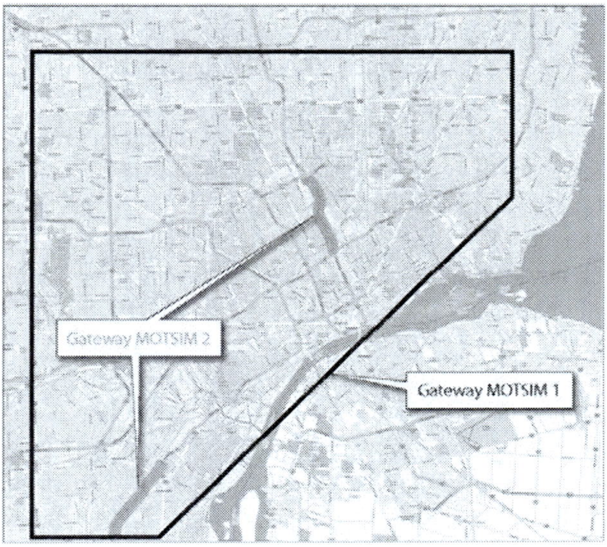

Figure 19. Ambassador Gateway Bridge MOTSIM Network Overview.

The results of the Gateway MOTSIM 1 analysis did not reveal any significant queuing or delay associated with the reconstruction project given the mitigation efforts that were to be implemented as well as the excess capacity available on other roadway links. However, during 2007 MDOT planned lane closures along section of I-75 to the north and south of the construction area that had not been included in the original analysis. The planning of the I-75 lane closures did not take into account the increased traffic due to the I-75/I-96 interchange reconstruction. The Gateway MOTSIM 2 analysis included the route diversions that would take place due to the I-75/I-96 reconstruction. The analysis revealed some areas of concern that lead MDOT officials to modify some of the planned mitigation efforts including travel advisories along I-94 and I-96, the major detour routes.

Gateway MOTSIM 3 began development in June of 2008 in response to feedback from stakeholders that lead MDOT management to alter some of the maintaining traffic ramp closures planned for July to October 2008. Also, traffic volume has been significantly lower since January of 2008. The results of MOTSIM 3 will point to further areas of mitigation and better performance measuring.

I-94 Rehab MOTSIM

The rehabilitation of I-94 in Detroit is an approximately six mile rehabilitation which will include the reconstruction of three interchanges (M-10, I-96 and I-75). The planning process has been occurring for the past 10 years at a cost of $11 million. Currently, the project is in the design stage of the decision-making process. MDOT refers to the current status as the Engineering Report (also known as pre-design or 10 percent of base plans) and has provided an initial construction cost of $1.3 billion. The purpose of the network simulation is to support the project in terms of constructability, staging, maintenance of traffic, cost estimates, financial plans, horizontal and vertical alignments and right-of-way planning. Currently, the team involved in the I-94 Rehabilitation project is using the I-94 Rehab MOTSIM to begin screening construction staging alternatives.

In addition to being another application of the SEMSIM data, the I-94 Rehab MOTSIM is the beginning of developing a new business process for MDOT whereby planners, traffic engineers and designers are all working together and incorporating the use of traffic models early in the decision-making process and using it to help answer many different types of questions. Documenting this process will provide a template for further process improvement.

I-75 Trade Corridor MOTSIM

The I-75 corridor between Detroit and Pontiac, Michigan is part of the truck network that supports critical supply chains linking automobile parts suppliers and manufacturers in Michigan and Ohio in the United States to each other and to the Province of Ontario in Canada. MDOT is currently planning for widening a 19-mile segment from outside Detroit to the City of Pontiac from three to four lanes. Currently in the planning stages (nearing completion of Engineering Report), the I-75 Trade Corridor MOTSIM will enable MDOT planners, engineers and designers to work together and test various scenarios to ensure mobility throughout the work zone, constructability is maintained and staging is adequate. The I-75 Trade Corridor MOTSIM will build upon the SEMSIM network, adding additional freeway links of I-696 and M-59, and Michigan surface arterial roads.

Findings

Process

Early in the development of the SEMSIM and Gateway MOTSIM, process issues immediately started appearing. With the then-current processes in MDOT, large network simulation could turn out to be a very resource intensive undertaking. Unless processes were developed to expedite model development and maintenance, network simulation might be unfeasible. And if each simulation model were limited to a single use or application, this cost might not be justified. The following describes three process issues that emerged.

Approach: Stand-alone, top-down, or bottom-up?

Before SEMSIM was conceived, all of MDOT's microsimulation models were project- specific, stand-alone models. They did not encompass or relate to the larger network, and they did not relate to each other. Thus, while each project had potentially profound consequences for the network, none of the models were capable of analyzing these. The technology of the time did not provide the needed plasticity for large, multi-purpose, multi- application models. Subsequently newer microsimulation software packages with cut-and-paste capability emerged and this advance allowed an ambitious project like SEMSIM to be conceived. For the first time, an existing model could be modified and reused by multiple applications. The technological problem appeared to have been solved.

Initially, the SEMSIM was conceived to become a complete top-down metropolitan area-wide or "master model", to be updated, and maintained on an ongoing basis, with a dedicated staff and budget, much like an MPO (Metropolitan Planning Organization) model. It soon became apparent that this approach would be resource intensive, and in any case, events were overtaking the process and MDOT's ability to implement this idea. What evolved by necessity was a bottom-up, incremental approach, whereby the SEMSIM and each subsequent derived, would serve as a core network for the next baby model. Instead of maintaining a "Master Model", each project would take whatever models had been developed to date, and update and customize them to its own needs. The costs of each customized model would be born by each project which used it. This leads to a problem of assuring uniformity between and among models, but this problem could be addressed by QA/QC or other institutional devices.

Data

Large scale, network simulation uses lots of data, especially traffic data. Initially, the fear was that there would not be enough data. It soon became apparent however, that the problem was not insufficient data, but rather poor or non-existent data processes. In a scramble to assemble system-wide data in a format that lent itself to mass data analysis and processing, several underlying process issues came to light. At the same time, MDOT had several tools or initiatives that held potential to resolve the problem, most notably, the Michigan Framework. The Framework was a unified GIS system that serves all agencies at all levels of government and even private industry. The Framework allowed every single directional link or structure, or any item that could be located in three-dimensional space, to be assigned a unique identifier (such as a physical reference number and milepoint). MDOT was in the process of inventorying and storing all its traffic and structure data (such as roads, ramps, bridges, traffic counts, bill boards, guardrails, and even ash trees) as Framework layers. This was to be an ideal platform for storing and retrieving traffic data and even coding the SEMSIM network.

The first issue was assembling large masses of freeway volume data from hundreds of stations across the system. The process of geo-coding traffic data collection sites to be Framework layers was accelerated to serve the SEMSIM. In addition, the project scope included the development of traffic analysis tools, such as would aid in cleansing and inputting traffic data and balancing in and out volumes along a corridor. A second issue was turning-movement traffic counts at signals. The traffic signal re-timing studies that were being performed were being collected by various consultants under different contracts, administered by engineers in the various Transportation Service Centers around the State. Not only were these traffic studies not centrally stored, they were not even known to a central data collection office. In fact, there was no central data collection office for the whole Department. A few quick phone calls to DOT's around the Country, confirmed that MDOT was not unique in this respect. Presently, MDOT is undergoing a systematic process improvement to rationalize and integrate all Departmental traffic data, and to geo- code it so it can be layered on the Michigan Framework. Finally, Michigan now has six inch resolution aerial photography of all its counties, in a projection that is consistent with the Framework. Thus, a Framework trunkline map and a "Sufficiency" file (with road segment characteristics data) can be layered on an aerial photo, with a zoom factor that allows pavement markings to be observed. These can be exported as bitmaps and imported into the simulation software, and network coding greatly expedited.

Organization

Simulation models and data collection are the easy part. They have to do with technology and software and numbers. The really difficult process issues have to do with the Organization itself. The best technology in the world can't solve a DOT's problems if its organizational processes are not compatible, or do not "fit", the new possibilities of the technology. For example, it took almost an entire generation, from 1970's to the 1990's before most government organizations fully adapted to the potential and promise of computer technology. Today, few could imagine an office or business without computer technology. They have transformed our processes, and at the same time, have pushed out our production possibilities curve. Likewise with simulation. Old simulation models lie in dusty files around MDOT, under- or unused, because they didn't get the right information to the right people at the right time. Much of this was due to the limited capabilities of the old technology. But it was also due to the fact that the old organizational structure hadn't caught up to the new possibilities. A lot more cross-organizational integration would be required. More organizational flexibility would be needed, so that processes that need to iteratively test scenarios can get traffic model updates as needed, and then pass output to downstream processes which may also need customized traffic model output.

In 2008, a concept for integrating MDOT's business processes to support an Operations- oriented environment was developed, called "Integrated Transportation Systems Operations and Management" or ITSOM.

Concomitantly, the Department was reorganized and a new Division of Operations was created. The emphasis throughout, is integration of tasks and functions across MDOT to support an integrated approach to Operations. The Division of Operations combines and extends previous functions such as ITS, Vehicle and Infrastructure Integration (VII), Work Zone Mobility and Safety, Traffic and Safety, and Maintenance. Strategic objectives and task forces charged with implementing these strategic objectives were established. Two of these objectives illustrate the new emphasis on cross-organizational integration and flexibility: 1. "Data Data Data" which would integrate data and data systems across the Department, and 2. "Integrate across the Organization", which includes, for instance, new channels and forums for communications and inter- divisional training. With implementation of ITSOM, MDOT will be positioned to fully harvest the potential of simulation to support all its Operations functions.

New Jersey Turnpike Authority and Rutgers University New Jersey Lane Closure Application[9]

Overview

The New Jersey Turnpike Authority, consisting of both the New Jersey Turnpike (150 miles) and Garden State Parkway (180 miles), funded the development of a modeling application that provides operations managers and engineers with information regarding queue and delays resulting from lane closure events. The tool, known as the Rutgers Interactive Lane Closure Application or RILCA is GIS-based and was developed by the Rutgers University Intelligent Transportation System Laboratory. The impetus for the development of tool was, in part, to put into electronic format hard-copy versions of traffic volume tables that operation managers would reference in order to approve or disapprove maintenance and contractor applications for lane closures related to roadway maintenance (guardrail repair, sign repair, etc.).

The lane closure tool was developed using the ArcGIS software package as the main development environment. Included in the GIS database is detailed geometric data for all of the roadway links for both the New Jersey Turnpike and Garden State Parkway including surrounding highways and local streets. Now, the RICLA tool provides operations managers, supervisors and engineers with the following features:

- **Traffic Demand**—Volume information on selected links at a given time period on any given date.
- **Roadway Data**—Link characteristics including number of lanes, AADT, milepost and link length.
- **Lane Closure Schedule**—A function that generates lane closure schedule for selected links based upon hourly volume data processed by the Rutgers Intelligent Transportation Systems (RITS) team.
- **Visualization**—A simple visualization function that shows the extent of expected queuing and delays as a result of lane closure and spill back onto upstream links in the form of link colors.
- **Cost Impact**—Integrated lane closure cost estimation function.

Today, two versions of the tool currently exist: one for the New Jersey Turnpike (called the Rutgers Interactive Lane Closure Application or RILCA) and one for the Garden State Parkway (called the Parkway Operations Lane Closure Application or POLCA). A third version is being developed for the New Jersey Department of Transportation.

Applications

The following applications are available in both RILCA and POLCA

- **Lane Closure Schedule**—A schedule for the minimum number of lanes to be open on a roadway section can be generated. These schedules are generated using the Annual Average Daily Traffic (AADT) data for the selected day. Lane closure schedules can be generated for a selected day or a week starting with a selected day for each month. An annual percentage increase in traffic can also be entered as an input to obtain the schedules for the future traffic demand.
- **Lane Closure Info**—The Lane Closure Info option is used to generate estimates of user delay and queuing resulting from a particular lane closure scenario. Outputs include Level of Service (based on the HCM procedure for work zone studies), minimum number of lanes to be open, hourly number of vehicles in queue, average hourly delay per vehicle, total hourly delay, and queue length.
- **Incident Analysis**— The Incident Analysis option is used to generate delays and queues resulting from an incident. It operates in similar fashion to the Lane Closure Function but includes the two additional

outputs of time of maximum queue length and time of queue clearance.

The New Jersey Turnpike Authority has not developed any specific polices or procedures regarding the use of the lane closure application tool. Currently, users of the tool are advised to use the results of the analysis as just one piece of the overall analysis and to also consider personal experience with lane closures and previous examples comparing estimated values to actual values.

Data

Data for the lane closure applications are routinely updated but are quite different for each version of the tool. For the RILCA version, hourly traffic volume data is generated from a combination of roadway sensor data and electronic toll collection (ETC) tag data. These data are combined to generate hourly traffic volume data which is entered into the GIS database every two months.

The POLCA version also use ETC tag data but this data is not as accurate as the New Jersey Turnpike since users only pay when they enter the parkway. The ETC tag data is combined with toll plaza hourly factors to estimate hourly traffic volumes for roadway links. These data are updated monthly, but the data is three months old due to the amount of processing.

Nova Scotia Department of Transportation & Public Works (Canada)—Reeves Street

Overview

Nova Scotia is one of Canada's Atlantic maritime provinces located on the east coast with a provincial population just under 1 million. Nova Scotia is composed of several islands and peninsulas with key commercial and population centers connected by the Trans Canada Highway. Port Hawkesbury is a key commercial location on Cape Breton Island just south of the Trans Canada Highway. The town of Port Hawkesbury includes a number of major industrial facilities representing companies such as Georgia-Pacific, Statia Terminals, USG, and the Sable Offshore Oil Company. The town also includes a number of residential communities.

In 2001, the intersection of Reeves Street and Trunk 4 in Port Hawkesbury, a location along a key access route to the Trans Canada

Highway, was slated to be upgraded. Both Reeves Street and Trunk 4 are generally 2-lane highways with certain sections upgraded to include designated turn lanes in built-up areas. Average Annual Daily Traffic (AADT) was on the order of 8,000 vehicles per day with approximately 10% of those being trucks. Truck traffic is concentrated during the daytime on weekdays, while trucks make up fewer than 5% of traffic volume on weekends. Most of the traffic volume at this location in Port Hawkesbury is through traffic continuing on to some of the industrial centers near the town or to points further north.

The reconstruction involved a major upgrade of the intersection including additional dedicated turn lanes to accommodate higher traffic volumes and to improve safety. In order for construction to take place, overall capacity of the intersection would be reduced due to narrow lanes widths and periodic lane closures. Construction was slated to take place only during daylight hours because of cost and safety concerns. However, it was also evident that any construction taking place during the day would have an impact on motorists, since the Reeves Street/Trunk 4 intersection carries a large amount of traffic. Therefore, in order to reduce the impact to motorists as much as possible, a sketch planning traffic modeling tool was used to test various construction phasing alternatives, including analyzing the possibility of a detour route, and to also provide the necessary data to justify the additional expense of night work.

Nova Scotia Department of Transportation & Public Works (Canada)
Reeves Street[10]

Work Zone Characteristics	
Transportation Analysis:	
Approach	Sketch-Planning
Modeling Tools	QuickZone
Work Zones:	
Type	Type III
Network Configuration	Isolated
Geographic Scale:	
Work Zone Size	Small
Analysis Area	Local

Work Zone Characteristics

Reeves Street intersection is a fairly basic design and consists of the signalized intersection and a detour route as shown in the figure below. Reeves Street ends at a "T" intersection with roads Trunk 4 and Trunk 4a to the left and right. The left turning movement continues on to Trunk 4 and carries a majority of the traffic volume. The right turning movement continues on to Trunk 4a and terminates shortly thereafter in large industrial area. In order to analyze the salient traffic movements, only the left turning movement headed north-east and the right turn movement headed south-west were modeled. This can be considered a corridor unto itself with the capacity limited primarily by the left turning movement lanes and signal phasing. The detour route consisted of Sydney Road which cut through a mostly residential section.

Detailed traffic data was available for this intersection from a traffic study which was carried out in 2000. The demand data, along with the signal timing plans, helped to determine the capacity of the intersection. Thus, the capacities of the various turning movements helped to determine the capacity of each link. Capacities under work zone conditions were estimated by the resident engineer from prior experience with flagging operations. The capacity and demand along the detour route was measured during the peak periods and on weekends by hand.

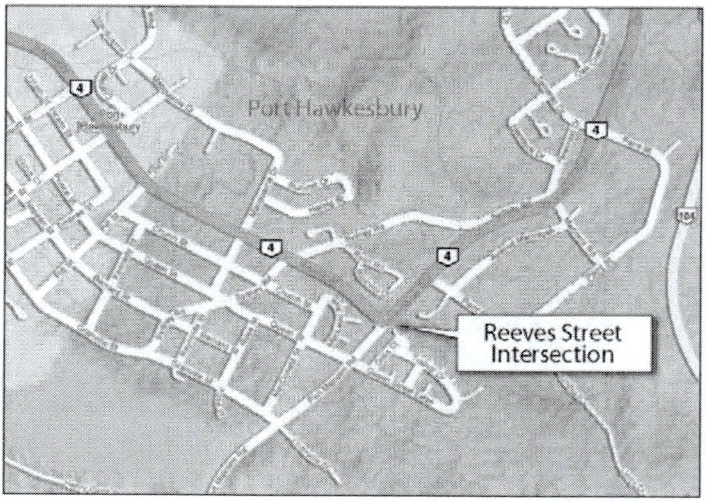

Figure 20. Reeves Street Intersection.

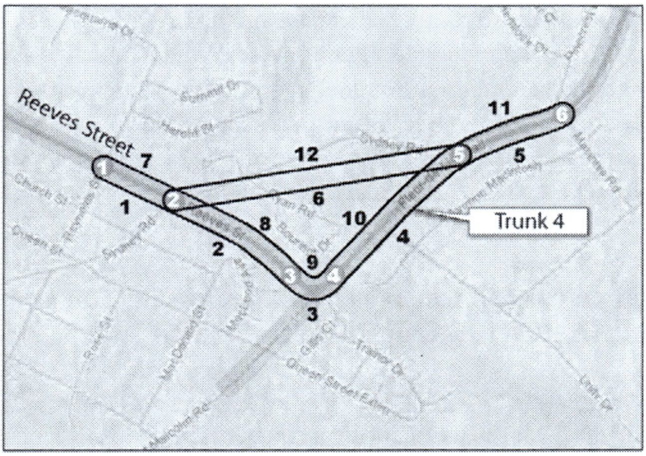

Figure 21. Reeves Street Network Model.

Application

The local engineers suspected that the reconstruction of the Reeves Street intersection would result in significant queuing and delay and that performing the construction at night could potentially mitigate many of these impacts. Initial results from traffic model proved these suspicions and showed an estimated maximum queue of 4.1 miles resulting in a 70 minute delay occurring on Friday evenings. Queuing and delay were also seen on Monday, Tuesday, Wednesday and Thursday but were approximately half the length and duration as seen on Friday evenings. Detour traffic was estimated at 10,000 vehicles per week. Knowing that the queuing, delay and detour volumes through the residential areas would be contentious issues with residents, the local engineers tested various construction phasing scenarios.

The first scenario had the work zone in place and operating with reduced capacity 24 hours a day, 7 days a week. Queuing during the day on Friday and Saturday were predicted to be especially severe. The engineers noted that queuing and delay did not form during the overnight hours on any day and tested a scenario that eliminated construction during daylight hours on Friday and Saturday (essentially shifting work to night). The results of this scenario cut in half the queuing and delay associated with the construction to 2.19 miles and 36.4 minutes respectively. In addition, the number of weekly vehicles on the detour route was reduced to 6,000, a 40% reduction.

Additional refinements in the work zone design, including tweaking the hours and days of operation, enabled the engineers to better schedule the construction activities so as to not impact motorists. Also, the Nova Scotia Department of Transportation & Public Works did not have a night-time construction policy and did not routinely approve projects for night work due to cost and safety concerns. Because of the results generated by traffic model, the decision was made to carry out some of the most disruptive phases of construction at the Reeves Street intersection at night.

Utah Department of Transportation & University of Utah—Evaluation of the I-15 Reconstruction Design-Build

Overview

In 1996, the Utah Department of Transportation began reconstructing a 17-mile segment of I-15 in Salt Lake City, Utah in preparation for the 2002 Winter Olympic Games. The reconstruction of I-15 was the culmination of nearly ten years worth of planning and analysis that resulted in the decision to widen I-15, construct HOV lanes and install a new a light rail system serving downtown Salt Lake City. In deciding to reconstruct I-15, the Utah Department of Transportation (UDOT) ultimately made the decision to move forward with a Design-Build construction method for two reasons: 1) the public strongly supported timely I-15 completion so as to minimize traffic congestion on alternate routes; and 2) Design-Build would complete reconstruction before the 2002 Winter Olympic Games began in Salt Lake City.

The decision to move forward with a Design-Build methodology for I-15 in Utah was the first of its kind for the state. Thus, a number of state procurement laws had to be changed in order to make it possible. In addition, UDOT had to restructure the manner in which they oversaw construction projects of this size and impact. Overall, the decision to use Design-Build was a leap-of-faith for UDOT with the understanding that the overall project would be constructed more quickly, include lower costs and not as greatly impact the traveling public.

Because the I-15 reconstruction was the first use of Design-Build for UDOT, an *ex ante* evaluation of the project was conducted by the Utah Traffic Lab at the University of Utah. The evaluation consisted of modeling three construction scenarios:

Utah Department of Transportation & University of Utah Evaluation of the I-15 Reconstruction Design-Build[11]

Work Zone Characteristics	
Transportation Analysis:	
Approach	Travel Demand Model
Modeling Tools	VISUM TP+
Work Zones:	
Type	Type I
Network Configuration	Network
Geographic Scale:	
Work Zone Size	Large
Analysis Area	Metropolitan

- **Design-Build** (aka Fast Track)—Innovative construction technique whereby time savings are gained as design and construction occur simultaneously. Cost savings occur from shorter construction duration and lower user delay and costs.
- **Traditional-Build**—More common design and construction method that typically includes lower capital costs but protracted construction duration and congestion.
- **No-Build** (aka Do Nothing)—Planned transportation improvements in the region would be implemented. Structural and pavement deficiencies on I-15 would be corrected.

The Utah Traffic Lab used the VISUM travel demand model to evaluate the three construction scenarios. Evaluations were based upon three criteria: user delay, I-15 corridor travel times and network congestion. In addition, an accident analysis was conducted to see the impact of the construction methods on safety. In the end, the evaluation of the reconstruction of I-15 in Salt Lake City led to the development of a construction impact modeling methodology (Figure1 below) that local jurisdictions can employ to justify the use of the Design-Build contracting. The methodology has been applied to five projects within Utah's Statewide Transportation Improvement Program with the recommendation that all five use Design-Build.

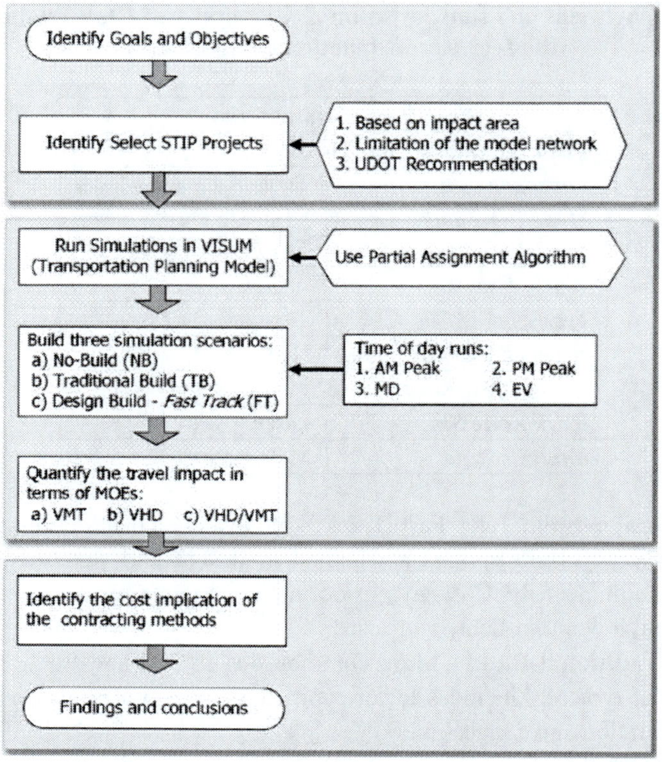

Figure 22. Construction Impact Modeling Methodology.

Overall, the evaluation of the Design-Build construction scenario estimated a savings of 60 million hours of user-delay over the Traditional-Build scenario representing approximately $600 million (2002 dollars) in savings to travelers (value of time = $10 per hour). In addition, the accident analysis estimated a reduction in vehicles crashes, injuries and fatalities of 287, 100 and 1 respectively. This is an estimated savings of $120 million (2002 dollars).

Model Selection

The study area for the evaluation comprised the entire Salt Lake Valley (Figure 23 below), about 500 square miles. The study area includes four major types of roadways: freeways, principal and minor arterials and collector roads (local and residential streets were excluded). Freeways include I-15, I-80 and I-215. In all, the area to be modeled included more than 5,000 roadway links.

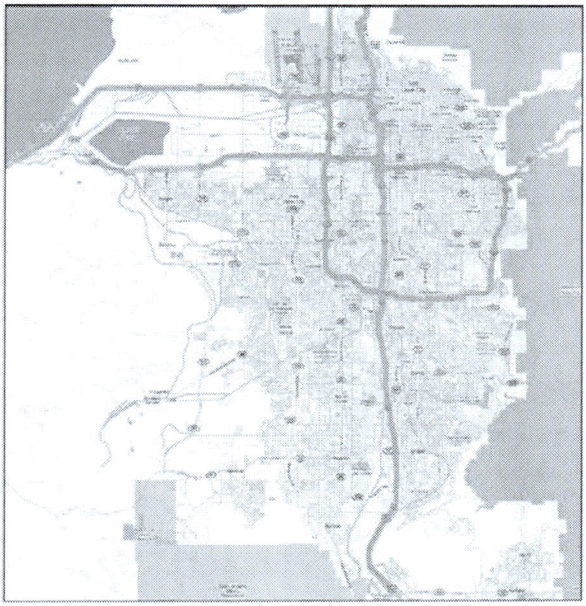

Figure 23. Salt Lake Valley Area.

In order to model such a large geographic area, the Utah Traffic Lab considered various transportation modeling software tools which included travel demand models (TransCAD, TP+, EMME/2, VISUM), mesoscopic simulation models (Integration) and microscopic simulation models (Paramics). Model selection was made based upon the following criteria:

- Quality of traffic assignment functionality
- Size of the network - number of nodes and links that can be handled
- Available traffic assignment routines
- Potential to export inputs/outputs to a microscopic simulation software package
- Number and variety of performance measures produced
- Price of the software (discounts, academic versions, technical support)
- User interface
- Peer reviews on the weaknesses and advantages of the software

VISUM was selected because it satisfied the given criteria. VISUM data is efficient as it can be directly exported to the VISSIM traffic simulation package giving the analyst an opportunity to use compatible microscopic simulation models for operational analyses. Practically speaking, Utah Traffic Lab owned a version of VISUM and the VISSIM model, keeping costs down. Also, the Wasatch Front Regional Council (the local metropolitan planning organization) modeled the region using the TP+ travel demand model which can be easily imported into VISUM. The VISUM Salt Lake Valley network is shown below in Figure 24.

Measures of Effectiveness

User Delay

User Delay was measured in terms of annual user delay and cumulative user delay. The values were calculated from VISUM output of Vehicle Hours of Delay (VHD) which is standard model output and is a region-wide measure of traffic in the system. VHD represents the difference between vehicle-hours on a link when it is congested and in free-flow conditions. Results (shown in Figure 25) indicate that the Design-Build scenario produces less user delay during the study period than both the Traditional-Build and No-Build.

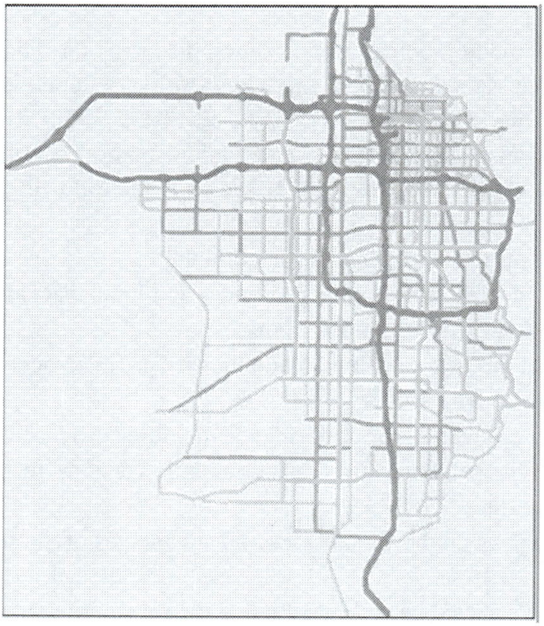

Figure 24. VISUM Salt Lake Valley Network.

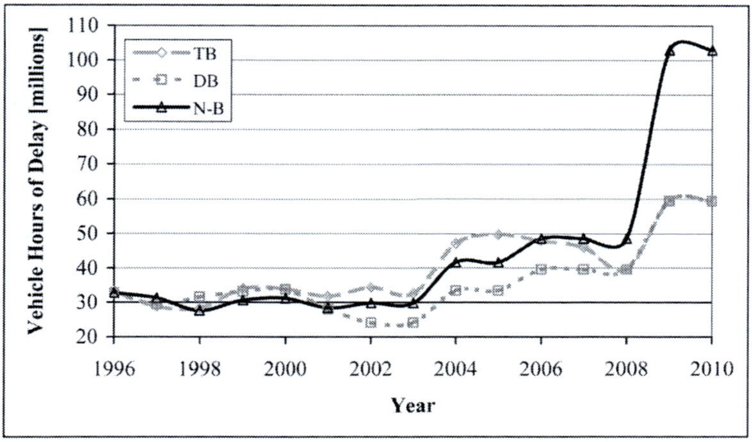

Figure 25. User Delay.

Travel Time

Travel Time was measured only on certain sections of I-15 using the route-search option in the VISUM software. This measure was used to evaluate the impact of different traffic loads during and after the I-15 reconstruction periods. Figure 26 shows that travel time are generally less for the Design-Build over the Traditional-Build and significantly less than the No-Build.

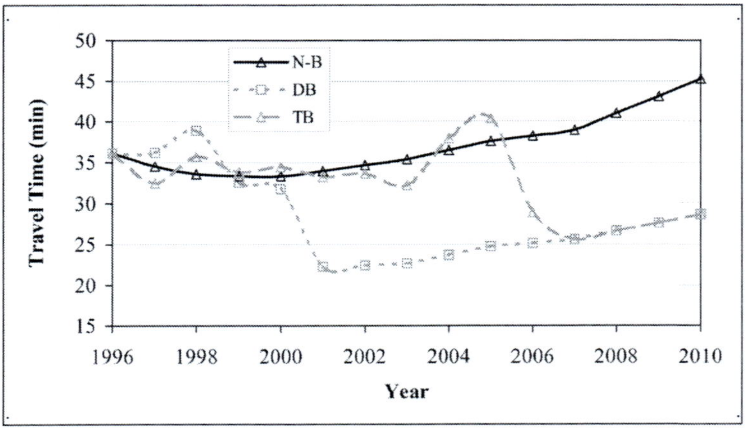

Figure 26. Travel Time.

Congestion

Congestion was measured in terms of volume-to-capacity ratios where saturation occurs with a value greater than 0.9 (1.0 represents volume equal to capacity). Conducting this part of the analysis was done off-line from VISUM using Microsoft Excel. Results are shown in Figure 27 and are difficult to interpret. Suffice it to say, the Design-Build produces more congestion early-on but less overall compared to both the Traditional-Build and No-Build.

Crash Analysis

An important component to the evaluation of the Design-Build scenario was the impact on accidents. In order to conduct a comparison among the three alternatives, a regression model was constructed which included a traveler exposure component represented by the vehicle-miles traveled (VMT) output from the VISUM model. Other regression components included work zone length, number of interchanges opened and congestion on the network. The regression model was calibrated to actual condition between 1996 and 2001 in order to predict accidents and accident rates for the three construction scenarios.

Results were expressed in terms of crash rate and total number of crashes. Figure 28 shows the calculated accident rates for the three construction alternatives for both highways and surface streets. Overall, the Design-Build scenario has a lower rate for the duration of the study period even though it is greater during certain periods. The total number of crashes during the study period is smallest for Design-Build (65.3 thousand) compared to both Traditional-Build (69.7 thousand) and No-Build (66.2 thousand).

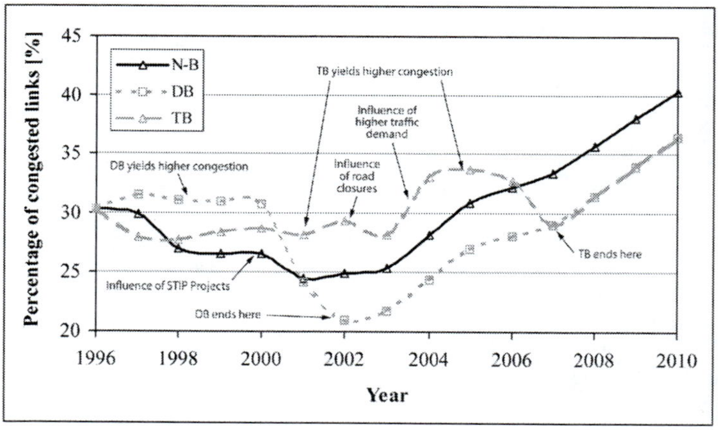

Figure 27. Congestion.

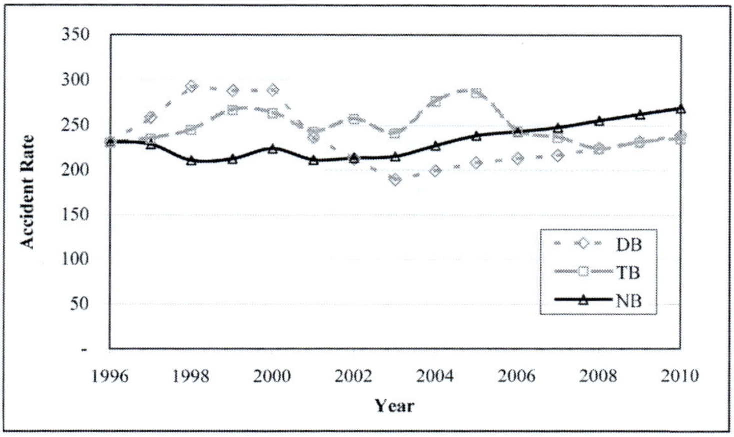

Figure 28. Crash Rates.

Wisconsin DOT—Work Zone Signal Optimization

Use of Signal Optimization Tools in Work Zone Traffic Analysis

Signal optimization tools such as Passer, Synchro/SimTraffic, and Transyt 7F have a variety of applications for work zone analysis, especially in urban and suburban environments. Broadly speaking, these applications can be grouped in three categories:

1. Preparing timing plans for temporary signals used to manage traffic within a construction site.
2. Adjusting signal timing on corridors that are directly impacted by construction.
3. Adjusting signal timing to improve progression on corridors that serve as alternate routes or detours around a work zone.

Temporary Signals. Figure 29 shows an example of the use of Synchro/SimTraffic to optimize the timing of a temporary traffic signal. In this case, two-way one-lane operation will be in effect during a bridge construction project (in other words, eastbound and westbound traffic will be sharing a single lane). Synchro's Ring/Barrier Editor was used to create a configuration that mimics the operation of the temporary signal by alternately sending eastbound and westbound traffic along the restricted section. Synchro's signal optimization algorithm was then used to establish a timing plan that minimizes traffic delays. The analysis also provides an indication of the extent of queuing on the approaches to the one-lane segment, which is useful in determining whether access to side roads will be blocked by queued traffic.

This method can also be used to evaluate the impact of work zone length on capacity and throughput for sites with two-way one-lane operations. As shown in Figure 30, the capacity of two-way one- lane sections is sensitive to the length of the restricted section. Therefore, in many cases there is a trade-off between what is convenient for construction operations and what is acceptable in terms of traffic impact.

Wisconsin DOT Work Zone Signal Optimization

Work Zone Characteristics	
Transportation Analysis:	
Approach	Traffic Signal Optimization
Modeling Tools	Synchro/SimTraffic
Work Zones:	
Type	Type I and IV
Network Configuration	Isolated
Geographic Scale:	
Work Zone Size	Small
Analysis Area	Local

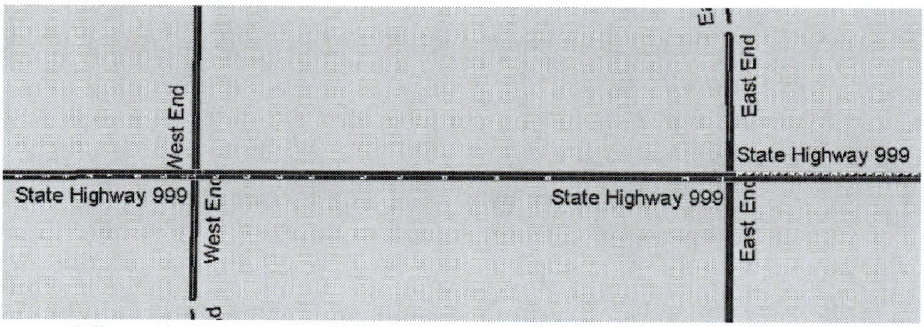

Source: Wisconsin Department of Transportation.

*Note the westbound vehicles queuing while eastbound traffic is allowed to proceed.

Figure 29. Synchro/SimTraffic Model of a Work Zone with Two-Way One-Lane Operation.

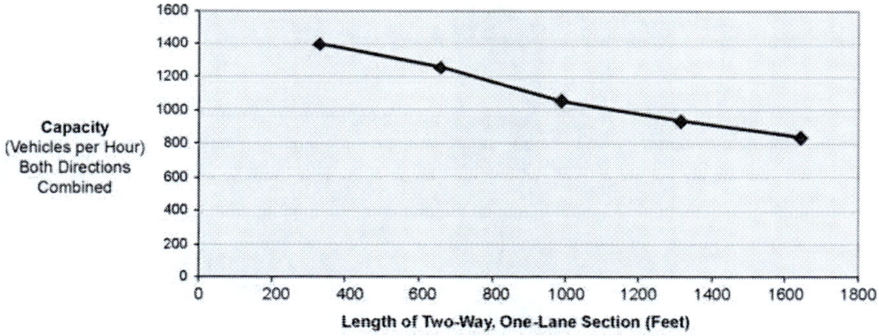

Source: UK Department for Transport, ***Safety at Street Works and Road Works***.

Figure 30. Capacity vs Length for Two-Way One-Lane Flagging Operations.

Adjusting Timing on Corridors Affected by Construction. Normally, signal timing plans are developed based on the assumption that all of the lanes that exist at each intersection will be available for traffic to use. This assumption may not be true during construction. For example, take an intersection where a two of the three lanes have been closed to traffic. In this case, all traffic is directed to use the right lane, severely impacting the capacity of the signalized intersection.

In such situations, to avoid excessive queuing and delay it may be necessary to make fundamental changes in the signal timing at individual intersections or along an entire corridor. In the example shown in the photo, it may be desirable to increase the cycle length to compensate for the fact that left, thru, and right turning vehicles are sharing a single lane. To maintain good traffic progression along the corridor, signal offsets may need to be adjusted to account for reduced travel speeds. In addition, temporary changes in access to business properties along the corridor may affect turn patterns, requiring adjustments in signal phasing and splits. The use of a signal optimization tool allows all of these variables to be addressed comprehensively.

Adjusting Timing on Parallel Routes. The Daniel Webster Hoan Memorial Bridge carries Interstate 794 over the Milwaukee River in Milwaukee, Wisconsin. As shown in Figure 31, on December 13, 2000 there was a structural failure on one span of the bridge. The failure required immediate lane closures, resulting diversion of all traffic to other routes.

Source: Unknown.

Figure 31. Structural failure on Daniel Webster Hoan Memorial Bridge, December 13, 2000.

To accommodate increased traffic on the arterial street that runs directly parallel to I-794, the City of Milwaukee used signal optimization tools to prepare a revised traffic signal timing plan for the Kinnickinnic Avenue/First Street corridor (WIS 32). The revised signal timing plan was implemented less than 48 hours after the incident occurred (and at minimal cost). It increased the green time allocated to north-south thru traffic, and reduced the amount of time allocated to side streets. The revised signal timing is believed to have been instrumental in reducing traffic delays and minimizing the overall impacts of the bridge failure and subsequent reconstruction activities.

Wisconsin Department of Transportation Work Zone Transportation Management Plans[12]

Overview

The Wisconsin DOT (WisDOT) has established guidelines on the development of Transportation Management Plans (TMP) as part of overall project delivery process during the design and construction stages. The formal process for developing a TMP is shown in Figure 1 below.

While the extent of the TMP depends upon the nature of the project (location, size, duration, etc.), WisDOT recommends that a systematic assessment of work zone impacts be conducted for construction projects that impact travel lanes. (Minor maintenance projects that do not affect roadway capacity, classified as Type 1 TMPs in Wisconsin, generally do not require an analysis). Development of the TMP is designed to mitigate, to the greatest extent possible, safety and mobility impacts to workers and the public.

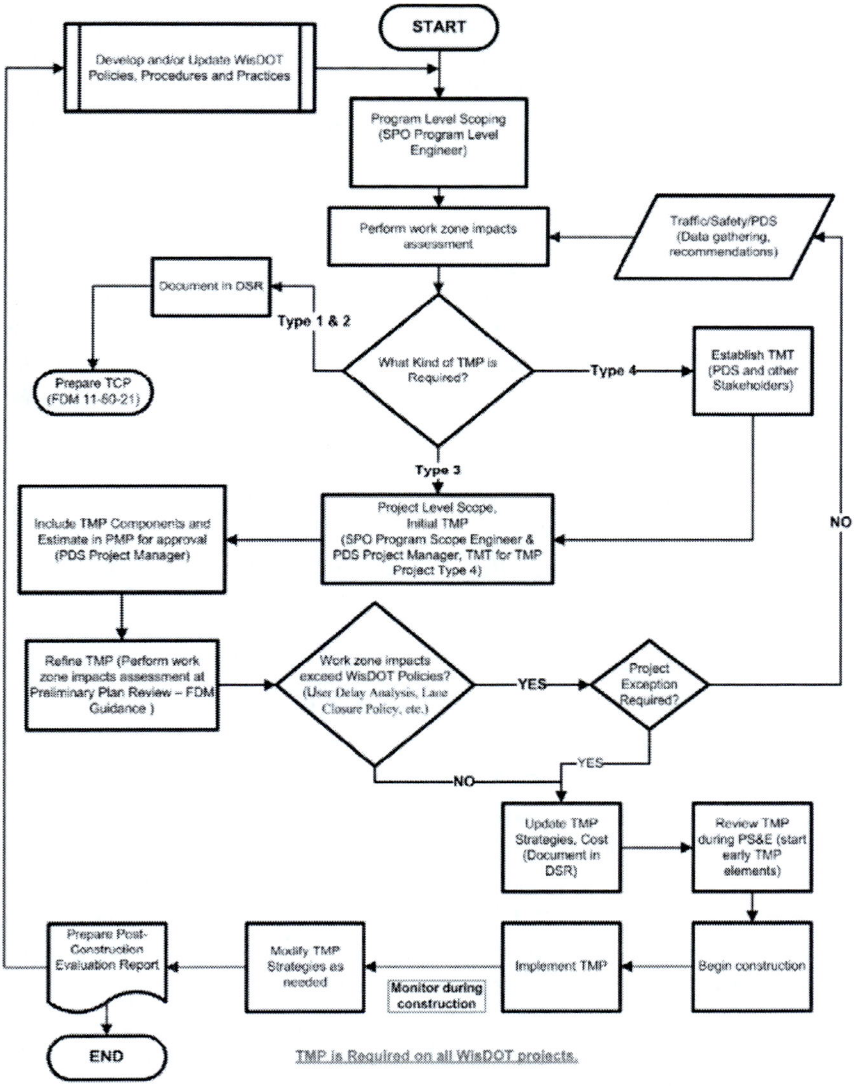

Figure 32. WisDOT Transportation Management Plan Developement Process.

In order to conduct the work zone impact assessment, WisDOT has formalized guidance used by regional offices in determining the extent to which work zone impacts need to be analyzed. This document is called the Lane Closure and Delay Guideline. With the guideline, WisDOT provides recommendations on the selection and use of traffic modeling tools for assessing the impact of various types of work zones.

A web-based database called the Lane Closure System (LCS) has been developed to handle the administrative process of approving lane closures. It is also used to keep track of lane closure status and will feed closure data to Wisconsin's 511 traveler information system.

Lane Closure and Delay Guidelines

In 2007 WisDOT updated their Facilities Development Manual to include guidelines for planning typical lane closures and methodologies for considering regularly occurring high volume periods with special considerations for holidays and planned special events. The lane closure guidelines establish a formal process which planners, engineers and operations managers follow in order to evaluate proposed lane closures. The process includes the following six steps (visually represented in Figure 33):

1. Determine route-specific maximum delay guideline and recommended lane closure times.
2. Estimate capacity under proposed lane closure.
3. Estimate hourly demand profile.
4. Estimate queues and delays using appropriate tools.
5. Identify appropriate mitigation strategies.
6. Plan and prepare for special conditions.

The guidelines include suggested procedures and methodologies to estimate the capacity of a roadway segment, determine traffic demand, and quantify resulting queues and delays using traffic volume data. Important to the discussion at hand is the recommendation of using traffic modeling tools with which step 4, Estimate queues and delays using appropriate tools, can be accomplished.

In addition to developing a formal procedure to systematically quantify queues and delays, WisDOT also developed a formal policy on acceptable delay values for construction projects on the Interstate system and other major multi-lane highways. The policy recommends no more than 15 minutes of delay above the normal travel time between key city nodes and within each city node. The guidelines suggest that a rough threshold estimate is when a queue length in excess of one mile is sustained for more than thirty minutes then the work zone is likely not meet the 15 minute delay criteria.

The following sections will provide more discussion on WisDOT's guidelines on the selection, use and implementation of various traffic modeling tools to be used for estimating queues and delays.

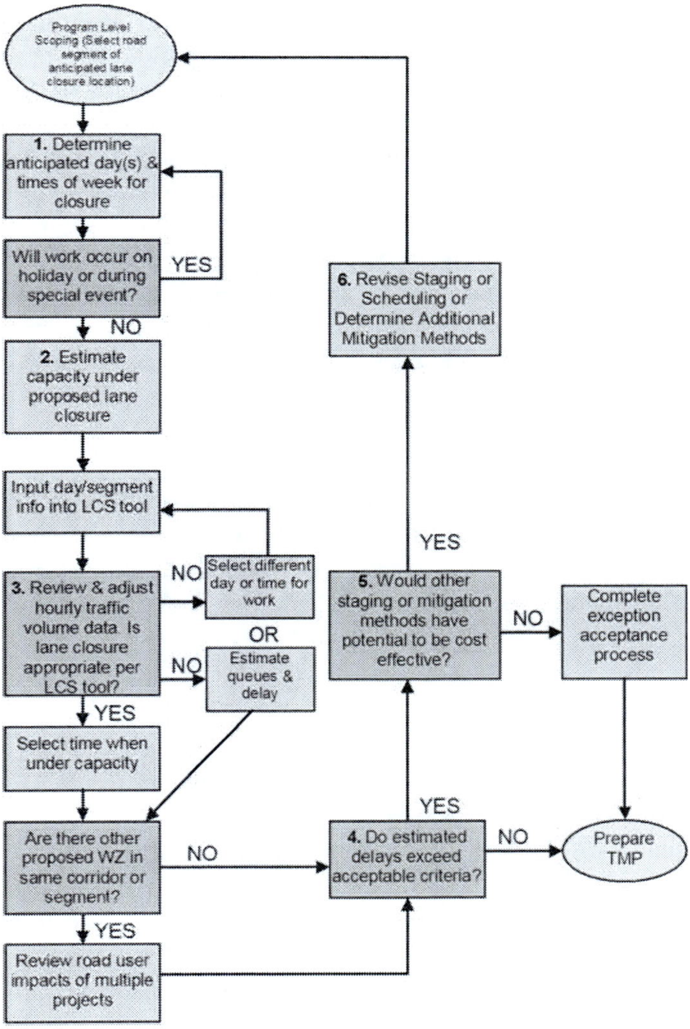

Figure 33. Lane Closure Analysis Process.

Queue and Delay Estimation Tool Recommendations

HCM Methodologies

- **Work Zone Type: 1, 2, ~3**
- **Network Configuration: Isolated**
- **Work Zone Size: Small**
- **Analysis Area Dimension: Local or Site**

WisDOT recommends the use of HCM-based traffic modeling tools for work zone Types 1 and 2, and perhaps certain Type 3 work zones. In order to support WisDOT engineers, WisDOT developed the Work Zone Capacity Analysis Tool (WZCAT). It is primarily intended to help predict delays and queues associated with short-term (daily) lane closures on rural freeways. The current version of WZCAT uses HCM methodology to calculate delay due to work zone operations based on two inputs.

One input is "work zone capacity" which is simply the capacity estimated for the work zone and the second input is demand which is traffic volume usually estimated from a single detector location upstream of the work zone. This tool is limited by the use of only a single work zone location, and it does not take into account the impact of heavy vehicles on queue estimations (though a revised version is being developed).

WZCAT calculates only the queues and delays associated with an over-capacity situation; currently the spreadsheet does not attempt to compute the delays associated with changes in travel speeds resulting from other characteristics of the work zone such as reductions in lane width and speed limits.

Sketch Planning Tools

- **Work Zone Type: 2, 3, ~4**
- **Network Configuration: Isolated, Pipe, Network (limited)**
- **Work Zone Size: Small, Medium, or Large**
- **Analysis Area Dimension: Metropolitan, Local, or Site**

WisDOT recommends the use of mode robust sketch planning tools (such as QuickZone or Quadro) for work zones of Types 2 and 3, perhaps certain Type 4. In Wisconsin, these tools are used primarily for analysis of projects on freeways and rural highways.

With the sketch planning tools, users can compare the impacts of multiple construction staging and phasing alternatives, work times, lane closures, traffic diversions, and various mitigation strategies.

Sketch planning tools require more data than HCM based methods. Typical requirements include route length, free flow speed, project duration, road closure times, and traffic volumes for primary and alternate routes. QuickZone requires creation of a network of links and nodes, while Quadro uses a simplified network consisting of two parallel links (one link represents the route that is under construction and the other represents an alternate or detour route).

In Wisconsin the additional data is relatively easy to obtain, and it is generally perceived that for projects of moderate complexity, the extra information provided by the software is worth the additional effort.

WisDOT has found that sketch planning tools are quite sensitive to assumptions about the capacity of the work zone. QuickZone requires capacity values to be supplied by the analyst. Quadro has built-in default capacity values for various types of work zones, which can be overridden if the analyst has suitable local data.

Wisconsin DOT advocates the use of sketch planning tools to help evaluate trade-offs between construction traffic management strategies, such as deciding whether the cost of implementing a more expensive strategy is justified by user delay cost savings.

Sketch planning tools are used to prepare a road user delay cost analysis that assigns a monetary value to delays caused by queues, reductions in speed through the work zone, and increases in travel time and vehicle operating costs associated with diversion to alternate routes. Since detours often involve sending traffic onto roads built to lower standards, the Quadro software also estimates crash probabilities and converts them to a monetary value.

Intersection Analysis Tools

- **Work Zone Type: 2, 3, 4**
- **Network Configuration: Network**
- **Work Zone Size: Small, Medium, or Large**
- **Analysis Area Dimension: Metropolitan, Local, or Site**

In general, sketch planning tools do not directly account for "control delays" associated with signalized intersections, roundabouts, or other traffic control devices. These delays can be significant for urban arterial projects and other locations where construction activities are likely to disrupt normal traffic patterns and redistribute traffic in the network. In these cases WisDOT recommends checking and updating signal timing plans using the estimated volumes that will be present during construction.

In Wisconsin this is usually done using a signal analysis and optimization tool such as Signal 2000, Synchro, or Transyt7F. These tools can also be used to evaluate the need for temporary signals. WisDOT uses Rodel to verify roundabout capacities.

Microscopic Simulation

- **Work Zone Type: 4**
- **Network Configuration: Generally Network**
- **Work Zone Size: Generally Medium, Large, or Very Large**
- **Analysis Area Dimension: Metropolitan, Local, or Site**

In Wisconsin, the Type 4 TMP designation generally applies to large, complex projects with regional impacts, such as freeway-to-freeway interchange reconstruction projects.

For these projects a microscopic traffic simulation program is often appropriate to determine the extent of queuing, user delays, and traffic redistribution in the network. Using microscopic simulation, individual vehicles are modeled in fine detail for the duration of their entire trip, providing traffic flow, travel time and congestion information, as well as potentially enabling the modeling of the interface between drivers and ITS. These programs provide a highly detailed analysisbut require detailed traffic and roadway data. Creating a properly calibrated model can take an extensive amount of time.

As a result, in Wisconsin microsimulation analysis to support the TMP is often integrated with constructability analysis and other traffic modeling done to verify and refine the overall design of the project.

To assist regional offices in conducting microscopic simulation analysis, WisDOT developed the *Microsimulation Model Scoping Worksheet* which is used by Bureau of Highway Operations to determine whether or not microscoping simulation is a) appropriate and b) technically feasible. The scoping worksheet is attached.

MICROSIMULATION MODEL SCOPING WORKSHEET

Version 0.01
19 Dec 2006

Name of Project ›	**Project ID** ›
Name of Requester ›	**Date of Request** ›
Geographic Extent of Model	
☐ West Limit ›	☐ South Limit ›
☐ East Limit ›	☐ North Limit ›
Roadways to be Included in the Model ☐ Freeways ›	☐ Principal Arterials ›
☐ Minor Arterials ›	☐ Collectors ›
☐ Local Streets ›	☐ Driveways ›
Roadways to be *Excluded* from the Model. › ›	
Briefly describe the existing traffic problems in the modeled area. › › ›	
Briefly describe any emerging/future issues that the model needs to address (such as new development). › › › ›	
How Much Detail is Required for Each Of These Roadway Elements?	
Major Roads ☐ Very Detailed ☐ Detailed ☐ Average ☐ Rough ☐ None	**Minor Roads** ☐ Very Detailed ☐ Detailed ☐ Average ☐ Rough ☐ None
Intersections ☐ Very Detailed ☐ Detailed ☐ Average ☐ Rough ☐ None	**Signal Timing & Phasing** ☐ Very Detailed ☐ Detailed ☐ Average ☐ Rough ☐ None
Roundabouts ☐ Very Detailed ☐ Detailed ☐ Average ☐ Rough ☐ None	**Ramp Meters** ☐ Very Detailed ☐ Detailed ☐ Average ☐ Rough ☐ None
Construction Stages ☐ Very Detailed ☐ Detailed ☐ Average ☐ Rough ☐ None	**Other (describe):** ›
Time Periods to Be Modeled:	
☐ AM Peak Period ›	☐ PM Peak Period ›
☐ Adjacent-to-Peak (Weekday) ›	☐ Off-Peak (Night) ›
☐ Saturday Shopping Traffic ›	☐ Other: ›
Is an Origin-Destination Study Available for this Area? ☐ No ☐ Yes (Describe): ›	**Are there previous models that can be adapted?** ›
Analysis Years (Horizon Year) ›	**Requested Completion Date** ›
Briefly Describe the Scenarios to Be Modeled and any Special Goals for the Model. › › ›	

Maryland State Highway Administration and the Virginia Department of Transportation—Woodrow Wilson Bridge Reconstruction

Overview

The replacement of the Woodrow Wilson Bridge (WWB), along Interstate 95 in the Metropolitan Washington, DC area, is an extremely large and complex construction project requiring the close coordination of numerous contractors and various state and local government agencies. The Woodrow Wilson Bridge carries upwards of 100,000 vehicles per day along Interstate 95 which spans the Potomac River. The Woodrow Wilson Bridge Project spans a 7.5 mile-long corridor extending from the MD 210 interchange in Maryland to Telegraph Road in Virginia, crossing over the Potomac River. The current 6-lane bridge is being replaced with a dual-span bridge that will more than double the number of traffic lanes. See Figure 34 for an overview of the construction site.

The management of the construction project primarily involves the Maryland State Highway Administration (MD-SHA) and the Virginia Department of Transportation (VDOT) who are responsible for overseeing daily operations of the construction site including scheduled lane closures. An important aspect to the replacement of the Woodrow Wilson Bridge is to maintain current roadway capacity during peak periods while construction takes place. This requires limited, if any, lane closures during peak periods and daylight hours. In general, any necessary lane closures were conducted at night to minimize the impact on drivers.

Due to the complexity of the project, the Woodrow Wilson Bridge Project is being completed under several phases over the span of more than a decade. In addition to the new bridge construction, a number of interchanges also have to be reconfigured both on the Virginia and Maryland sides to join to the new bridge alignment. With several jurisdictions involved, the long duration of the project, and 100,000 vehicles per day utilizing the bridge, traffic management during construction of the bridge and interchanges was a major concern. There were several types of work zones implemented throughout this project with which several different work zone modeling tools were used including the following:

- ***Lane Closure Analysis:*** **CLV**
- ***Roadway Operations Analysis:*** **CORSIM**
- ***Roadway Closure Analysis:*** **QuickZone**

Maryland State Highway Administration and the Virginia Department of Transportation Woodrow Wilson Bridge Reconstruction

Work Zone Characteristics	
Transportation Analysis:	
Approach	Sketch-Planning Mesoscale Microscale
Modeling Tools	QuickZone HCM CLV VISSIM CORSIM
Work Zones:	
Type	Type I
Network Configuration	Network
Geographic Scale:	
Work Zone Size	Small
Analysis Area	Metropolitan

Work Zone Characteristics

The Woodrow Wilson Bridge reconstruction project affected several miles of major roads and interchanges in two states including I-95/I-495 (Capital Beltway) from Maryland to Virginia, I-295, MD 414, and MD-210 in Maryland, and Telegraph Road, Route 1, and Washington Street in Virginia. To keep traffic moving throughout the longevity of the project, the construction work for the Woodrow Wilson Bridge project had to be completed in many steps instead of one large work zone. Work zone plans had to be developed for each step of this construction project and therefore, the Woodrow Wilson Bridge Project work zones are categorized as the most severe, Type I, work zones.

The Woodrow Wilson Bridge team faced many challenges in developing work zone plans, implementing detours, and maintaining traffic patterns while minimizing delays. In addition, the possibility of traffic delays and the visibility of this project made it imperative to provide accurate traffic information to commuters using the bridge during its reconstruction. The complex nature of the work zones on this project required extensive analysis of traffic impacts caused by work zones and traffic detours. The Woodrow Wilson Bridge team used several different work zone analysis tools to provide insight regarding traffic impacts and also predict delays to the public. Figure 34 shows the different work zone analysis tools used in this project.

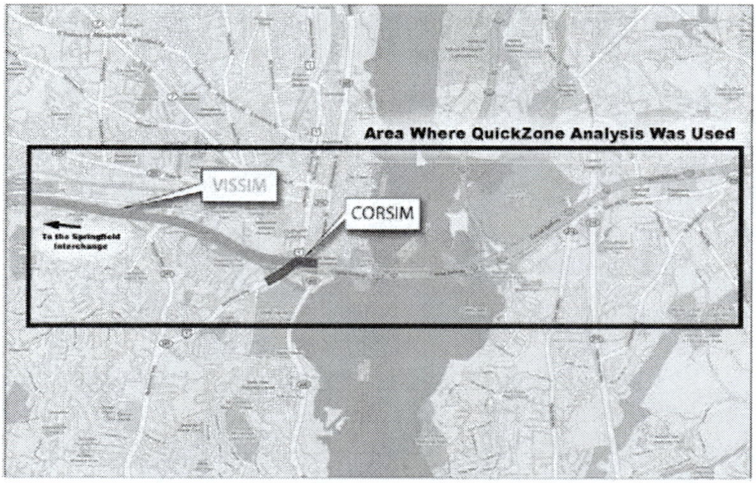

Figure 34. Analysis tools used for the Woodrow Wilson Bridge Reconstruction Project.

Transportation Management Plan Strategies

Traffic engineers on this project had the challenge of dealing with designs and maintenance of traffic (MOT) plans from two different jurisdictions. In addition, Virginia and Maryland had different philosophies for their MOT plans. The MOT plans for the Maryland side were very detailed and utilized models to evaluate their MOT phasing for some of their contracts including Synchro, Critical Lane Volume (CLV) and Highway Capacity Software (HCS). In addition, they used CLV to obtain hourly and 15 minutes traffic volumes during the construction phase. Conversely, the MOT plans for the Virginia side were much more fluid and changed significantly during the construction phase. They used both Synchro and QuickZone analysis during the construction phase to evaluate their work zone impacts.

There were numerous work zone plans required for the Woodrow Wilson Bridge Project varying between a Type I and Type III work zone. For example, the first use of QuickZone was for a Type III work zone associated with the reconstruction of the Maryland 210 interchange. There were three Type 1 work zones occurring throughout the project which involved major traffic switches and disruption to the traveling public. The first was a full closure of the bridge during overnight hours in the spring of 2006. The second and third involved two major traffic switches during the summers of 2005 and 2006. The following will provide an overview of each of the four work zones mentioned previously.

- ***Fall 2001 MD 210/I-295 Interchange Reconstruction (QuickZone)***—In the fall of 2001, a contractor for the Maryland State Highway Administration was in the process of constructing one of many new bridges as part of the replacement and refurbishment of the Woodrow Wilson Bridge at the MD 210/I-295/I-95 interchange just east of the Potomac River shoreline (Figure 35). The construction involved the demolition of existing bridges and the construction of new bridges. The plan included closing lanes during the overnight hours (midnight to 4:00am) and two temporary openings in the median barrier in order to divert traffic safely around the construction area while construction was taking place. This work zone plan configuration was scheduled to take between 4 and 6 months to complete.

  The need for QuickZone arose when the contractor determined that to more efficiently utilize available resources, a four-hour window of construction would not be sufficient. The original work zone plan had lane closures occurring between midnight and 4:00am and the contractor estimated it would take 1.5 hours for work zone set up and take down, leaving 2.5 hours of actual production time. The analysis required a quick turn-around of multiple scenarios of when lane closures could take place without severely impacting motorists so that the contractor could extend lane closure durations.
- ***Summer 2005 US 1 Traffic Switch (CORSIM)***—During Summer 2005, the US Route 1 Interchange contract was completed and included new ramp connections to the bridge. A traffic switch was needed to move from the old ramp connections to the new alignment as seen in Figure 36. This was a major undertaking and detailed traffic information was needed to assess the impacts to traffic. The design of the new alignment was complex and included a flyover ramp from southbound Route 1 to the northbound ramp of I-95. The engineers used CORSIM to analyze the switch because there were ramp alignment questions that could be answered with this software. The analysis took approximately 4 months (from late 2004 to early 2005). They used count data for the ramps that were performed in 2003 and mainline count data that were obtained from cameras mounted at the site.
- ***Spring 2006 MD 210 Girder Placement (CLV Analysis)***—In the spring of 2006, a contractor for the Maryland State Highway Administration was in the process of replacing the MD 210 Bridge

that goes over the I-495 Beltway. As part of this work, the contractor needed to erect very large steel girders for the MD 210 Bridge. The contractor could not place the girders while traffic was moving because the work would be done above the Beltway (I-495). The original plan included 15 minute lane closures for the placement of these girders but due to their large size, it was clear that the work could not be completed in these short time spans. Instead, all lanes of the Beltway (I-495) in both directions approaching the Woodrow Wilson Bridge were temporarily closed overnight for several consecutive nights. The traffic engineers on this project developed a plan to detour traffic during the full closure as seen in Figure 37. It was very important that the Woodrow Wilson Bridge team analyze the impacts to the capacity of traffic based on the detour route. They needed to find out what the volume needs would be during the closure and therefore used CLV analysis to obtain the critical lane volumes and figure out how the detour would affect capacity.

- ***Summer 2006 Traffic Switch (QuickZone)***—In the summer of 2006, the first of the two new WWB bridges was completed. A major traffic switch was performed on both the inner and outer loops of the beltway, as shown in Figure 38 and Figure 39, respectively, where all of the traffic needed to be switched from the old bridge to the new bridge alignment so that the old bridge could be demolished. There was pressure to provide an accurate assessment of the traffic impacts and relay those impacts to the public. The analysis was performed using QuickZone. QuickZone was chosen in part because it was very important to know accurate queues and delays due to construction and related those numbers to the WWB traffic call center. In addition, QuickZone was chosen because the switch would result in higher capacity and therefore it was expected that conditions would improve due to the switch.

Data

Data collection was very important to the traffic analysts working on this project. Although data availability was scarce in the beginning, specifically existing volume data in Maryland, as the project continued, the analysts were able to obtain better data that was essential in the analysis of traffic impacts. These data include daily real time data on the beltway provided by Mobility Technologies, 24-hour intersection counts and tube counts taken for Virginia and Maryland, ramp data as recent as June 2007 and the entire network of Synchro files for Virginia provided by VDOT (NOVA district)

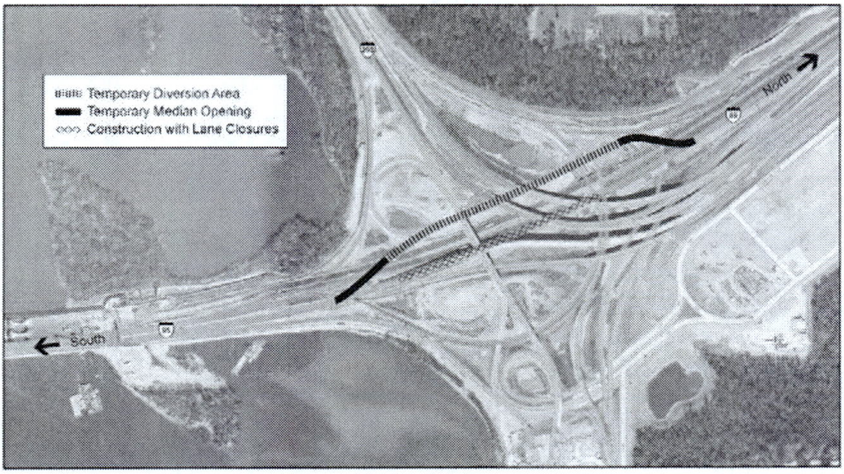

Figure 35. MD 210/I-295 Interchange Reconstruction.

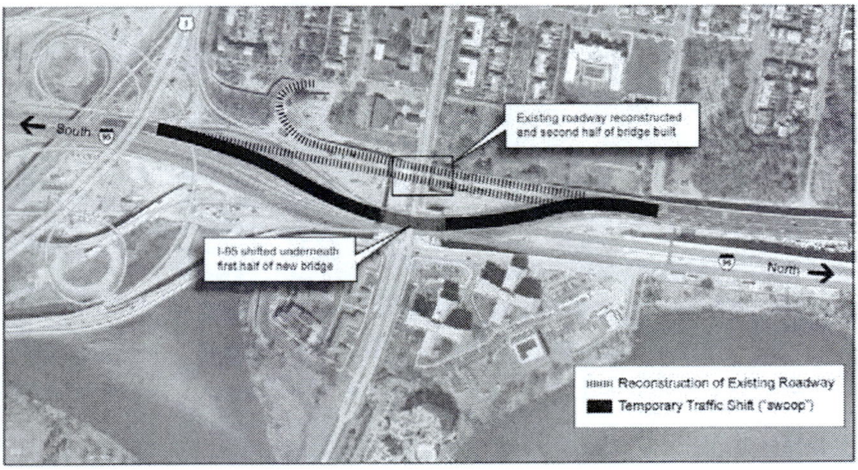

Figure 36. US 1 Interchange Traffic Switch Using CORSIM.

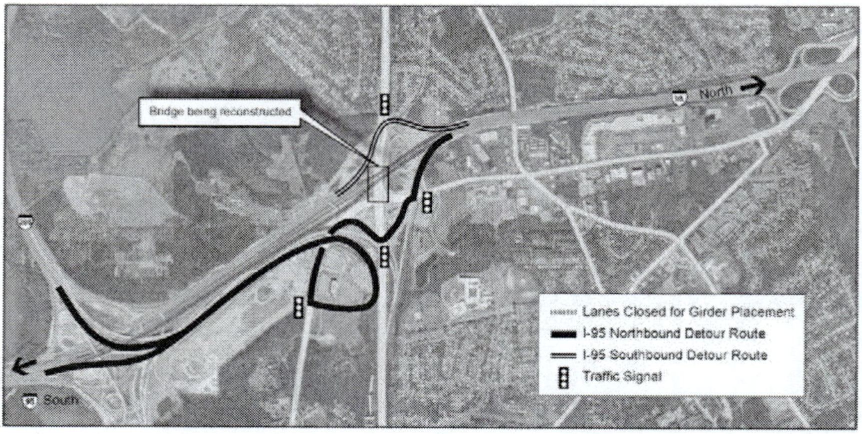

Figure 37. MD 210 Girder Placement Using CLV.

Figure 38. 2006 Inner Loop Traffic Switch Using QuickZone.

Demand data for the inner-loop was acquired from the various MD-SHA permanent traffic count stations set up prior to the construction site and available at the Maryland Roads web site (www. marylandroads.com). Demand data for the outer-loop was made available through special traffic counts conducted by the VDOT on the Virginia approach to the Woodrow Wilson Bridge. Lane capacities, lengths, and free-flow speeds were gathered from historical data.

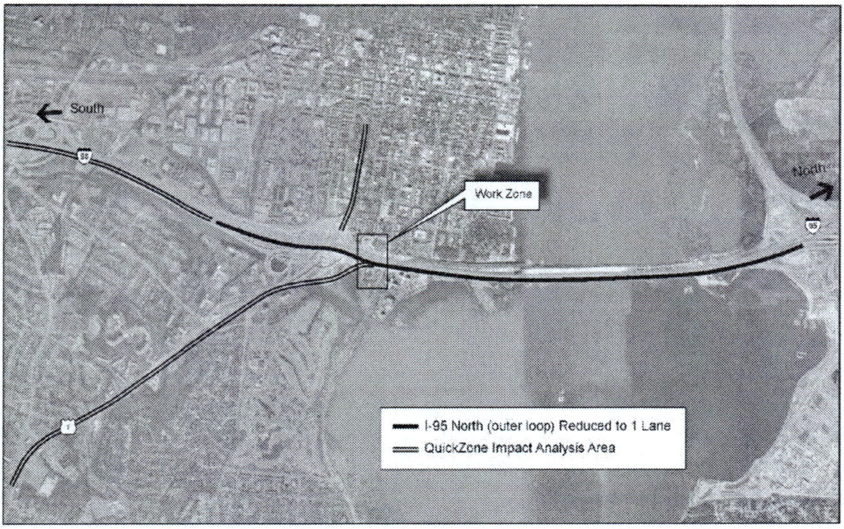

Figure 39. 2006 Outer Loop Traffic Switch Using QuickZone.

Agency Resources

Although the management of the construction project primarily involved MD-SHA and VDOT, the joint vengture General Engineering Consultants (GEC) was hired to oversee the construction. This joint venture included Parsons Brinkerhoff, URS Corporation, Rummel Klepper and Kahl (RK&K) and several sub-consultants including 8 DBEs. GEC had on-site technical expertise with modeling software and therefore much of the simpler work zone modeling that was done periodically, such as QuickZone, was performed in house.

Work Zone Modeling Tools

In addition to the major work zones discussed previously in this case study, there were other work zone modeling opportunities throughout the project and a range of modeling tools was used to support decision-making. QuickZone was used extensively, approximately 3 to 4 times per year, to understand impacts of the open span of the bridge and discuss the impacts with the Port Authority. The analysts used QuickZone for mainline analysis but chose other software for portions of the analysis where turning movements were important. The QuickZone software is a sketch planning tool that was "developed to help state and local transportation agencies better understand and consider the impacts of work zones as they plan, design, and implement their highway projects". This software can help estimate work zone delay and user costs, quantify corridor delays from capacity decreases in work zones, identify the impact of delay on different construction plan alternatives and estimate the impact of delay when developing mitigation strategies such as alternate routes, detours, lane closures.

Critical Lane Volume Analysis (CLV) is a planning tool that was used by the Maryland side to analyze level of service at individual intersections. CLV identifies levels of congestion at signalized intersections and it is useful because the results are easy to understand and to report to the general public. In the WWB project, hourly and 15 minute traffic volumes were provided using CLV.

Highway Capacity Software (HCS) is an operational tool that was used to model the expected added travel time that will be generated from the each construction zone. The outcome of this analysis can determine Level of Service (LOS) of the facility.

On the Virginia side, Synchro was used as a traffic signal optimization tool to understand the work zone impacts. Synchro's outcomes include LOS, signal phasing improvements for alternate routes and detours during construction, and travel delay estimations.

For the more complex modeling, microscopic simulation tools were used including VISSIM and CORSIM. Both of these software packages can simulate the movement of individual vehicles using car-following and lane change theories. For VISSIM, a base network is set up to run test simulations under different scenarios during construction. It can evaluate delays of the multiple scenarios using variable speed limits. CORSIM outputs maximum and average queue lengths, determines LOS and provides estimations of delays in work zones.

Yosemite National Park—Yosemite Village Roadway Reconstruction

Overview

Yosemite National Park in California is one of the most popular national park destinations in the nation, averaging more than 40,000 visitors through its entrance gates each day throughout the year. One of the primary entrance destinations for park visitors is Yosemite Village, the primary hub of activity within the park and home to the Valley Visitors Center, a wide variety of lodging and dining options, trail heads, and other visitor services.

The shape of Yosemite Valley Figure 40 makes access to Yosemite Village scenic for the park visitor but quite limiting for a traffic manager. Given the steep terrain around the valley, the only roadways into and out of Yosemite Village are Northside and Southside Drives running along the Merced River which flows through the center of the valley. Both facilities are two-lane, one-way facilities with stop-controlled intersections along their length at two bridge crossings.

These two key valley roadways are scheduled for a significant repaving and rehabilitation project scheduled to start in 2006 but subsequently delayed through funding and legal issues to this day (August 2008). The Federal Lands Highway Division (FLH) of the FHWA is responsible for the planning, design and construction phases of the project, working in conjunction with the National Park Service to minimize impact on park visitors and the environment while cost-effectively conducting the needed roadwork.

Concern regarding significant delays in the construction phase led FLH staff and NPS personnel to consider a range of phasing and staging alternatives. This concern is nontrivial given current (no roadwork) traffic conditions, where weekend congestion and delays are already a recurring event during peak travel months.

One alternative considered was an alternating full closure plan where work on Southside Drive could be conducted quickly while all inbound/outbound traffic would be directed onto Northside Drive (temporarily configured to support 2-way traffic). In a second phase, Northside Drive would be closed and all traffic diverted onto Southside Drive. The advantage of this alternative was that the project could be completed faster (one season) and more efficiently at a lower cost. The disadvantage of such an approach was that capacity reductions from the roadwork had to be in place around the clock, and could not be timed to avoid weekly and daily peaks in travel demand.

A second alternative was to pursue project planning under a more traditional approach where one lane of each facility would be repaved while the other remained open to traffic. This approach would allow for work to be suspended during peak demand hours but would be less efficient to conduct, lengthening the project duration to two seasons and incurring additional costs.

The original role of modeling the work zone impacts in the Yosemite project was to identify the likely travel delays expected under the two alternatives, allowing FLH and NPS staff to make an informed choice between the two, trading off road user delay against project cost. As the case study progressed, however, the traffic model became integral in the incremental refinement of a phasing and staging plan combining advantageous aspects of both alternatives.

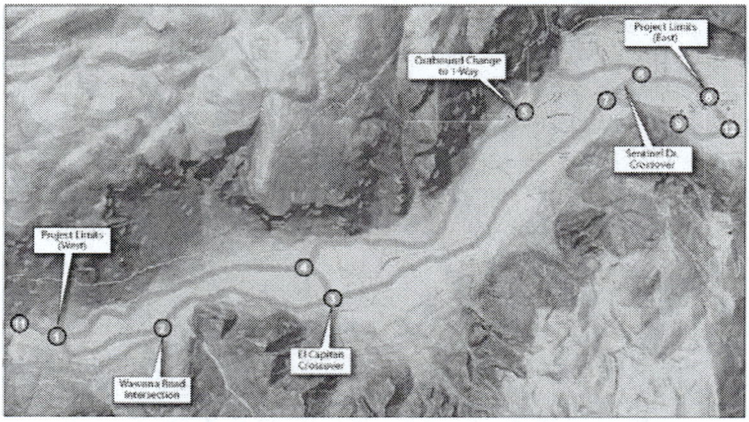

Figure 40. Yosemite Case Study Area.

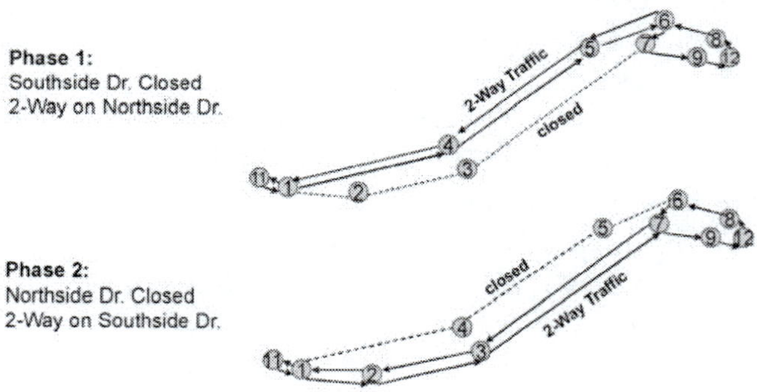

Figure 41. Yosemite Traffic Modeling Networks.

Network Design and Modeling Approach

Network design for each alternative was different given the significant changes in geometry and traffic flow associated with the full closure components of the single-season alternative. The first alternative was coded using two networks (Figure 41). In the first phase, links representing Southside Drive are removed from the network and new links representing inbound operations on Northside Drive are added. The reverse is true for phase 2, where links representing Northside Drive are removed from the network and new links added to Southside Drive for outbound operations. These networks were run as separate files in traffic modeling software and results were combined external to the model. The second alternative was coded using a single network for all project phases.

A key element for the work zone analysis was obtaining accurate travel demand. Travel demand data were assembled from a number of sources and then refined through two short-term data collection activities. A first-cut distribution of hourly and daily travel demand factors were obtained from a 1998 traffic study. Monthly variations in travel demand were obtained from park entrance station data. Finally, two-short term collection activities (one two-week collection activity in June 2004 and one one-week collection activity in August 2004) were conducted to refine hourly and daily distributions and to establish a rate of travel demand growth from 1998-2004. Given that recurrent weekend congestion had worsened over the period, data were collected to identify when and by how much visitors had shifted departure or arrival times to avoid congested periods since 1998 (Figure 42). The supplementary data collection effort in 2004 was conducted using NuMetrics Hi- Star portable traffic counters, a commonly utilized technology within FLH.

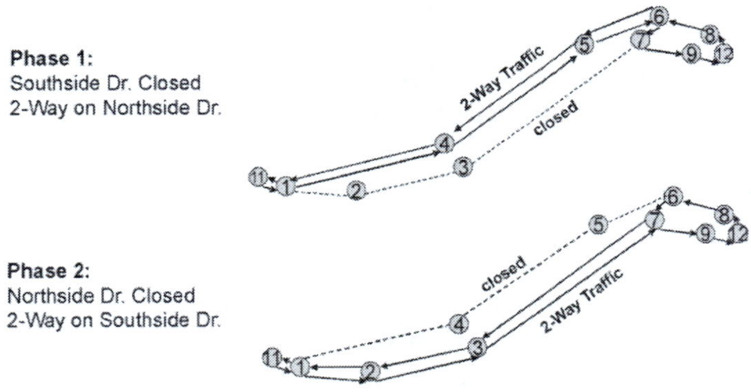

Figure 41. Yosemite Traffic Modeling Networks.

Traffic control operations modeled included lane closures, flagger operations, and full closures. Capacity of flagger operations were estimated external to the model and then input for affected links.

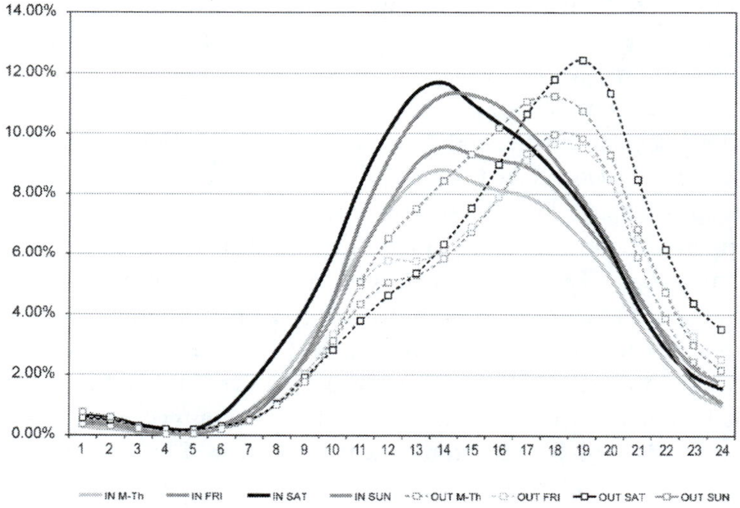

Figure 42. 2004 Hourly Demand Pattern by Day of Week (from HI-STAR).

Application

The single-season alternative, although cost-saving and shorter in total duration, was predicted to generate long and unacceptable delays for park visitors, particularly on weekend afternoons during summer months. The two-season alternative, when no-work hours had been refined by additional work zone analysis, produced no more than 10 minutes of additional visitor delay. The differential in road user delay between the single-season and two-season alternative was too large to justify the reduced cost of the single-season alternative.

In discussing the results, however, it became clear that the full-closure elements of the single-season alternative could be viable if the delay during the peak months of July, August and September could be avoided. In response, FLH staff developed a hybrid third alternative plan that combined full closure activity during relatively low-demand months (March-June, October-November), and traditional one-lane paving operations in the peak summer months.

This third alternative plan had the advantage of recouping most of the cost savings of the single- season approach with significantly lower travel delays. Delay was not eliminated, however. In June, outbound delays were predicted to approach 30 minutes for outbound traffic on Sunday afternoons. Likewise, in October, inbound delays on weekends could approach 60 minutes.

Zion National Park—Entrance Booth Reconstruction

Overview

Designated a National Park in 1919, Zion National Park is Utah's oldest national park. Zion canyon features soaring towers and monoliths that suggest a quiet grandeur. Zion is also known for its incredible slot canyons, including "The Narrows," which attract hikers from around the world. With nearly three million visitors per year, Zion is Utah's most popular National park. Entrance fees are $20 per vehicle and $10 per person arriving on foot.

There are two major entrances to Zion National Park, a south and an east entrance. A third entrance, located on the west side of the park, provides access only to Zion's Kolob Canyon. The South Entrance is the larger and most frequently-used of the three entrances. The South Entrance is on Utah Route 9 about 60 miles south from Cedar City, via I-15 and Utah Route 17 (see Figure 44 below). The East Entrance is on Utah Route 9, 12 miles east of Mt. Carmel Junction, at U.S. Route 89. The park contains over 12 miles of road. To ease traffic congestion within the park, a shuttle system is available to take visitors to the most popular areas. A shuttle bus service is also available from the town of Springdale, just outside of the park. Shuttles operate from April through October; during that period private automobiles are not allowed on the 6.5 mile stretch of road in Zion Canyon. The shuttles provide the only access to marquee attractions like the Great White Throne, the Watchman, the Grotto, Angels Landing, Weeping Rock and the Temple of Sinawava. Automobiles are allowed on other park roads, including all of Hwy 9, which provides access to the lower part of the park, the Tunnel and the East Entrance/Checkerboard Mesa area. Approximately 90% of park visitors utilize the south visitor's entrance. The entrance includes 2 visitor lanes and 1 employee lane, controlled by a radio-frequency tag system. Employees are able to process approximately 240 visitor vehicles an hour of which 50% are cash transactions, 40% credit card and 10% National Park Pass. 75% of the vehicles entering the park are passenger cars/ trucks and 25% are RVs. During the peak season, recurring queue can extend as much as ¼ mile from the entrance. In 2004, a major rehabilitation of the main road through Zion, beginning at the south visitor's entrance and extending into Zion Canyon was scheduled to take place. This included widening and structural repairs of certain sections of the existing road and milling/paving of the entire 7 mile stretch. A major concern of the National Park Service was the impact to visitors coming to Zion National Park through the town of Springdale. Traffic congestion on roads inside the park was not a concern since visitors are required to use the free shuttle bus service. Significant queuing and delay at the south visitor's entrance, where recurrent queues were already present, was of major concern to park administrators. The original work zone plan called for shutting down one visitor entrance lane at a time for construction. The National Park Service did not want construction to cause a queue to form that extended into the town of Springdale, approximately 1/2 mile from the south visitor's entrance. A queue of this length would not only impact traffic in the town, but employees getting to the park and the operation of the shuttle bus service from the town to the park. The work zone analysis and use of the traffic modeling tool was used to estimate the length of queue and number of vehicles in queue if one of the two visitor entrance lanes were to be closed for construction. This was conducted for the peak tourist months of June, July, August, September and October.

Zion National Park Entrance Booth Reconstruction[14]

Work Zone Characteristics	
Transportation Analysis:	
Approach	Sketch-Planning
Modeling Tools	QuickZone
Work Zones:	
Type	Type III
Network Configuration	Isolated
Geographic Scale:	
Work Zone Size	Small
Analysis Area	Local

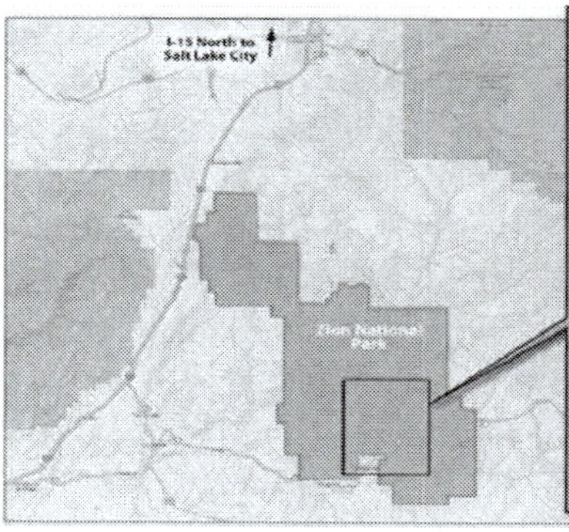

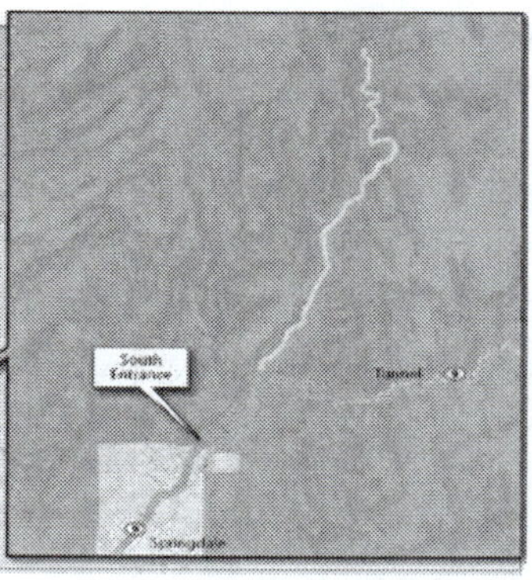

Figure 43. Zion National Park Overview. Figure 44. Zion National Park.

Work Zone Characteristics

The Zion National Park work zone is simple in design and can be considered an isolated work zone with no interaction of surrounding infrastructure. For the purposes of this work zone analysis, only the south visitor's entrance was modeled since there was no concern about queuing or delay along the seven mile stretch of road within the park.

A critical element regarding this work zone was estimating the capacity of the south visitor's entrance. This entailed calculating the number of vehicles that can be processed in an hour and is similar to estimating the capacity of a signalized intersection or toll plaza. Work zone capacity was calculated using average transaction time per vehicle and a breakdown of average transaction types per day. It was calibrated against historical queuing data to arrive at an overall facility capacity. Hourly counts were generated from a simple traffic study conducted in April 2002 for an average day both during the week and on the weekend. Seasonality was taken into account by using April as a baseline point and scaling up demand based upon historical knowledge for the months of June, July, August, September and October.

Construction at the south visitor's entrance was just one component of the overall construction project and was estimated to take between two to three weeks. The use of the traffic model was being used to estimate queuing impacts if construction were to occur during the months of June, July, August, September or October and to determine which month would cause the least amount of impact.

Application

The roadway construction at Zion National Park was planned to only occur Monday through Friday and during daylight hours. This was done to avoid impacting the larger crowds visiting on the weekends and disturbing those visitors camping during the evening and overnight hours. As a result, the work zone analysis focused upon the weekdays. Baseline queuing was calibrated against actual demand seen in the field during the month of April. Once the calibration was complete, traffic model was run where capacity of the south visitor's entrance was reduced by 50% (the equivalent of one lane of traffic entering the park).

The results of the work zone analysis indicated that the queue will impact the town of Springdale in each of the five months analyzed:

- **June:** .58 mile queue, 463 vehicles in queue
- **July:** .66 mile queue, 513 vehicles in queue

- **August:** .66 mile queue, 513 vehicles in queue
- **September:** .58 mile queue, 461 vehicles in queue
- **October:** .45 mile queue, 361 vehicles in queue

The order of magnitude for delay was around 300 minutes or 5 hours with the queue beginning to form around 9 AM, peaking at 3 PM and dissipating by 9 PM. Clearly, people will not wait 5 hours to get into the park. The key data point for this analysis was the estimated number of vehicles in queue and whether that queue will impact the town of Springdale.

Results of the Zion work zone analysis provided the local engineers with the necessary data to reevaluate the construction phasing. After seeing the results engineers knew that the current construction phasing could not take place as originally designed and began to brainstorm on various alternatives including opening up temporary entrance booths and shifting construction to the early evening hours to not coincide with the peak demand. In the end, the results of the work zone analysis provided the engineers with necessary evidence to require the reconstruction to take place at night.

References

[1] Federal Highway Administration, *Work Zone Modeling and Simulation: A Guide for Decision-Makers*. 2008.

[2] Federal Highway Administration, *Traffic Analysis Toolbox, Volume 1: Traffic Analysis Tools Primer*. 2004.

[3] Federal Highway Administration, *Work Zone Impacts Assessment: An Approach to Assess and Manage Work Zone Safety and Mobility Impacts of Road Projects*. 2006.

[4] Federal Highway Administration, *Developing and Implementing Transportation Management* Plans for Work Zones. 2005.

End Notes

[1] The assumption at this point is that a work zone analysis will be conducted. Section 4.1, Deciding to Analyze or Not Analyze, provides a detailed discussion related to the opportunities and risks associated with conducting a work zone analysis.

[2] The discussion of Transportation Management Plans is based upon the classification developed by FHWA in the report *Developing and Implementing Transportation Management Plans*

for Work Zones available at http://ops.fhwa.dot.gov/wz/resources/publications/trans_mgmt_plans/index.htm. A detailed discussion of each individual strategy discussed in the following sub-sections is available in the FHWA TMP document.

[3] A detailed discussion on the precision of the results is provided in Section 2.5.2, Precision of Results.

[4] This case study was adapted from the reports: *I-15 Pavement Reconstruction Project: Construction Operational Analysis* prepared by Cambridge Systematics, June 7, 2007; *Pre-Construction Analysis and Construction Staging Plan for The I-15 Ontario Pavement Reconstruction Project* prepared by EBL Consulting and RBF Consulting, October 2006; and *TAT Volume VII: Predicting Performance with Traffic Analysis Tools: Case Studies* to be published in 2008.

[5] Total Cost = (RUC/3) + Cost Analysis

[6] This case study was adapted from the report *FLH-QuickZone Case Studies: The Application of FLH-QuickZone in Six Federal Lands Projects* available from Federal Lands Highway Division.

[7] The EIS also reported that without the presence of major roadwork overall park visitation was expected to increase by 0.6 percent. The estimated decrease in visitation of 6.4 percent due to major roadwork is detailed in the EIS for Alternative 3 (Preferred Alternative).

[8] This case study was adapted from the LCAP User Guide available at http://marylandroads.com/safety/workzone.asp.

[9] This case study references the Rutgers Interactive Lane Closure for Work Zone Planning Manual prepared by the Rutgers University Intelligent Transportation Systems Laboratory (November 2007 edition).

[10] This case study was adapted from the report *The Application of QuickZone in Eight Common Construction Projects* available from the FHWA Office of Operations.

[11] This case study was adapted from the reports Evaluating Design-Build vs. Traditional Contracting Methods for STIP Projects: An Assessment of Travel Impact & Delay Cost (July 2004) and I-15 Reconstruction In Ogden, Utah: Evaluation of Various Traffic Maintenance Plans (July 2005) prepared by Dr. Peter Martin at the University of Utah.

[12] This case study references the document Guidelines for Developing Work Zone Transportation Management Plans (January 2008) prepared by the Wisconsin Department of Transportation.

[13] This case study was adapted from the report *FLH-QuickZone Case Studies: The Application of FLH-QuickZone in Six Federal Lands Projects* available from Federal Lands Highway Division.

[14] This case study was adapted from the report *FLH-QuickZone Case Studies: The Application of FLH-QuickZone in Six Federal Lands Projects* available from Federal Lands Highway Division.

INDEX

D

E

F

G

H

I

J

L

M

N

O

P

R

S

T